D0350047

Nature's Compass

Books in the *SCIENCE ESSENTIALS* series bring cutting-edge science to a general audience. The series provides the foundation for a better understanding of the scientific and technical advances changing our world. In each volume, a prominent scientist—chosen by an advisory board of National Academy of Science members—conveys in clear prose the fundamental knowledge underlying a rapidly evolving field of scientific endeavor.

The Great Brain Debate: Nature or Nurture,
by John Dowling

Memory: The Key to Consciousness,
by Richard F. Thompson and Stephen Madigan

The Faces of Terrorism: Social and Psychological Dimensions,
by Neil J. Smelser

The Mystery of the Missing Antimatter,
by Helen R. Quinn and Yossi Nir

The Long Thaw: How Humans Are Changing the Next 100,000 Years of Earth's Climate, by David Archer

The Medea Hypothesis: Is Life on Earth Ultimately Self-Destructive?
by Peter Ward

How to Find a Habitable Planet,
by James Kasting

The Little Book of String Theory,
by Steven S. Gubser

Enhancing Evolution: The Ethical Case for Making Better People,
by John Harris

Nature's Compass: The Mystery of Animal Navigation,
by James L. Gould and Carol Grant Gould

Nature's Compass

The Mystery of Animal Navigation

James L. Gould

Carol Grant Gould

PRINCETON UNIVERSITY PRESS
Princeton and Oxford

Published by Princeton University Press, 41 William Street, Princeton, New Jersey 08540
In the United Kingdom: Princeton University Press, 6 Oxford Street, Woodstock, Oxfordshire OX20 1TW

Jacket photograph: Canadian geese migration in sunset, © Alaska Stock Images/National Geographic Stock

press.princeton.edu

Library of Congress Cataloging-in-Publication Data

Gould, James L., 1945–
Nature's compass : the mystery of animal navigation / James L. Gould, Carol Grant Gould.
p. cm.—(Science essentials)
Includes bibliographical references and index.
ISBN 978-0-691-14045-2 (hardback : alk. paper) 1. Animal navigation. I. Gould, Carol Grant. II. Title.
QL782.G68 2012
591.56′8—dc23 2011035086

British Library Cataloging-in-Publication Data is available

This book has been composed in in Minion and Myriad

Printed on acid-free paper. ∞

Printed in the United States of America

10 9 8 7 6 5 4 3 2 1

For Verity,
whose loft may be found at
42°06′54.72″N
71°06′43.59″W

Contents

Preface *ix*

Acknowledgments *xiii*

1 Navigating—Problems and Strategies *1*

2 When and Where *19*

3 A Matter of Time *35*

4 Insect Compasses *69*

5 Vertebrate Compasses *117*

6 Piloting and Inertial Navigation *155*

7 The Map Sense *185*

8 Migration and the Future: Conservation and Extinction *227*

Bibliography *245*

Illustration Credits *281*

Index *289*

Preface

. . . and they spend their winters upon the moon.

This was the deduction of the "person of learning and piety" who published a treatise in 1703 on

> the probable solution of this question: whence come the stork and the turtle, and the crane and the swallow, when they know and observe the appointed time of their coming—or where those birds do probably make their recess and abode, which are absent from our climate at some certain times and seasons of the year?

The idea that migrants might overwinter on the moon may have fallen out of fashion, but not our sense of awe at the seemingly effortless way many animals come and go as they choose. Migrating animals have always been a mystery, appearing unexpectedly and vanishing again, en route from an unidentified home to an unknown destination—a living allegory of human existence. As the eighth-century scholar and historian the Venerable Bede (731) wrote in his *Ecclesiastical History of the English People*, "The life of man is but a moment of existence."

> Consider the swift flight of a sparrow through the room in which we sit at supper in winter, while darkness and a snow-

> storm prevail outside. The sparrow, flying in at one open window and immediately out at another, vanishes from our sight after experiencing a few seconds of warmth and light. So too the life of man is but a brief moment; of what went before, or what is to follow, we are utterly ignorant.

Some of that age-old sense of mystery remains, but our increasing knowledge has only served to deepen our awe at the phenomenon of migration. How incredible, that the tiny hummingbirds at our feeders have journeyed north 1500 miles for the brief summer, while fragile monarch butterflies are preparing themselves for a 2000-mile trip south in the fall. And yet, for them and myriad other species, it is perfectly routine. Most animals for which even local navigation is a common and crucial activity are performing feats far beyond anything humans can manage without specialized instruments, equipment, and training. What they know innately about the sun and sky could fill volumes. The still more remarkable ability of many species to deduce their global position to within about a mile is, for all we knew to the contrary even just a few years ago, magic.

There is no one answer to the mysteries of animal navigation. Orientation and migration are conundrums that have been only haltingly solved. The challenge has been difficult because mechanisms vary not only between species, but within the same animal in different contexts and at different ages. Creatures are using the sun and stars, polarized light and color gradients, endogenous timers (daily, tidal, lunar, and annual), landmark memory and cognitive maps, magnetic fields, extrapolated gradients, and more. They confound study by using cues redundantly, so that if we remove one source of information they simply switch to a backup, making a shambles of our attempts to understand their abilities. For some, their navigation system is an integral part of a rich code of communication.

This is a particularly opportune time for a book on animal navigation and migration. The increasing number of highly technical monographs and edited compilations fail to tell the story to the wider world. The last serious review for the well-read nonspecialist audience is more than two decades old. Since then a crescendo of work has resolved many of the debates over animal compasses and maps, redefined our understanding of how animals use time, and decoded the critical roles of early rearing, juvenile experience, and ongoing adult recalibration.

We have chosen to focus most of our attention on a subset of well-studied species. In our lab at Princeton we have studied the navigation behavior of honey bees and homing pigeons extensively. Sea turtles and migratory birds also are excellent models of animal navigation systems. Thanks to their combination of spectacular behavior and experimental convenience, these creatures account for the vast majority of research on navigation. Humans too come into the narrative, though often as an example of what not to do. The past quarter century has witnessed a general and ever-increasing pattern of disruption in migratory paths and behavior—a disruption largely attributable to human activities and global climate change. The twin threats of habitat loss and climate destabilization lead many researchers to ask whether the elegant programming that enables migration might now be leading migrating animals to oblivion. Perversely, a puzzling imprecision in the migration programs of many species may prove their best hope for dodging the bullet of extinction. And for certain high-risk species, our increasing understanding of animal navigation could provide the key to effective conservation efforts.

James L. Gould
Carol Grant Gould
Princeton, NJ
August 2011

Acknowledgments

It is our great pleasure to thank the several people who have helped us in our efforts. For general inspiration we can single out Ken Able, Don Griffin, Bill Keeton, Joe Kirschvink, Charlie Walcott, David Wilcove, and John Bonner. For repeated encouragement to synthesize the literature in the field, we thank Geoffrey North of *Current Biology*. We particularly appreciate the brilliant job Dimitri Karetnikov has done with the figures. We thank Barbara Clauson for her meticulous editing, and Alison Kalett for her important help in finely balancing the level and emphasis.

Nature's Compass

Chapter 1

Navigating—Problems and Strategies

■ Devil Birds of the Atlantic

It was a navigator's worst nightmare. Shortly after midnight on a cloudy September night in the middle of the North Atlantic, the ship was suddenly attacked. Out of nowhere the screeching and wailing ghosts of long-dead sailors swept through the rigging, terrifying the superstitious seamen and drowning the captain's shouted orders. The panicking crew knew instantly that they had trespassed on the infamous Isles of Devils, haunted by the souls of the thousands of crewmen who had perished on the treacherous shoals. But by the navigator's reckoning, they were far to the west of the legendary death trap.

Within minutes, though, over the din came the fatal sound of waves breaking on a lee shore to the east. Turning the ship and piling on sail was their final mistake: the keel ran hard into one of the lethal coral reefs that ring the islands. The ship sank almost immediately, joining countless others that had suffered the same fate in that graveyard of the Atlantic that is the Bermudas. What had gone wrong?

Long-distance navigation is a life or death challenge for many nonhuman animals as well; the difference is that *they* know what they are doing. Few sights are more impressive to earthbound people than an isolated formation of geese passing overhead on their way to distant summer or wintering grounds. Theirs is not a journey on a wing and a prayer: all but first-year birds have a detailed map of the route in their brains, complete with remembered landmarks for piloting. After dark in the spring and fall literally billions of songbirds traverse the skies each night unseen, often to destinations hundreds or thousands of miles away. Unlike waterfowl, these passerines are using multiple compasses and a mystical GPS sense to find their way. At least 30 species would have been passing overhead on that fatal night in the Atlantic, maintaining their steady course for 2300 miles from Nova Scotia to South America.

Animals were on the move in the surrounding waters of the North Atlantic as well, racing against time to reach new habitats as the seasons changed. Humpback whales migrate past Bermuda on thousand-mile journeys with maps and compasses adapted to the gloom of the ocean. Many fish and sea turtles are similarly equipped for equally monumental seasonal redeployments. Below them on the seafloor spiny lobsters prance in tandem lines on arduous journeys, knowing their location in the cold darkness to within a few feet.

On a more local scale, honey bees and many other insects on the Bermuda islands, as elsewhere, commute scores of times each day from home to sources of food, water, or building materials, using a series of backup compasses and learned landmarks. Relative to their small body size and myopic vision, these trips are nearly as epic as those of geese, and each journey is a life-or-death event. Nesting birds on this mid-Atlantic refuge log many hundreds of miles shuttling back and forth first to collect nest material and then food for their young; getting lost would mean starvation for the next generation. Mice do much the same on the ground, but must employ a different set of cues and processing strategies more suited to running mazes than finding distant continents. Monarch butterflies flutter 2000 miles south from the United States and Canada to a remote mountain peak in Mexico, orienting by the ever-moving sun and a mysterious sense of location.

Faced with what is to us an alien task in an unforgiving world, humans stand in awe of the judgment and precision with which animals use cues—often undetectable by us—that are frequently ambiguous and ephemeral. We must depend on luck as much as talent, trying with clumsy approximations to replicate the compass sense that animals use innately to work out the rules of piloting and mapping, and to pinpoint our position on the globe without the seemingly magical combination of sensory abilities and inborn processing circuits that for other species comes as standard equipment. The navigators whose ships came to grief on Bermuda's reefs were, in fact, employing an amalgamation of generally reliable animal strategies. What went wrong for that ill-fated ship beset by devils?

When Fernandino deVerar set sail in his well-armed, two-masted, 300-ton Portuguese merchant ship *San Antonio* in 1621, he understood the risks. He had sailed that spring from Cadiz in southwestern Spain for Cartagena in what is now northern Colombia with a load of goods for the American colonies. Like all human navigators of the time he was unable to judge his east–west position (the longitude) when out of sight of charted land, so his ship had sailed south with the northern coast of Africa clearly visible to larboard (the left, or port, side of the ship). He had abandoned this piloting strategy once his collection of instruments, charts, tables, and measurements of the sun's elevation combined to tell him he was 12° north of the equator—roughly the latitude of his destination 3000 miles on the other side of the Atlantic. Then deVerar had turned west and, using a magnetic compass, sailed along this latitude until he reached the Caribbean. This vector strategy is similar to the way many migrant songbirds navigate

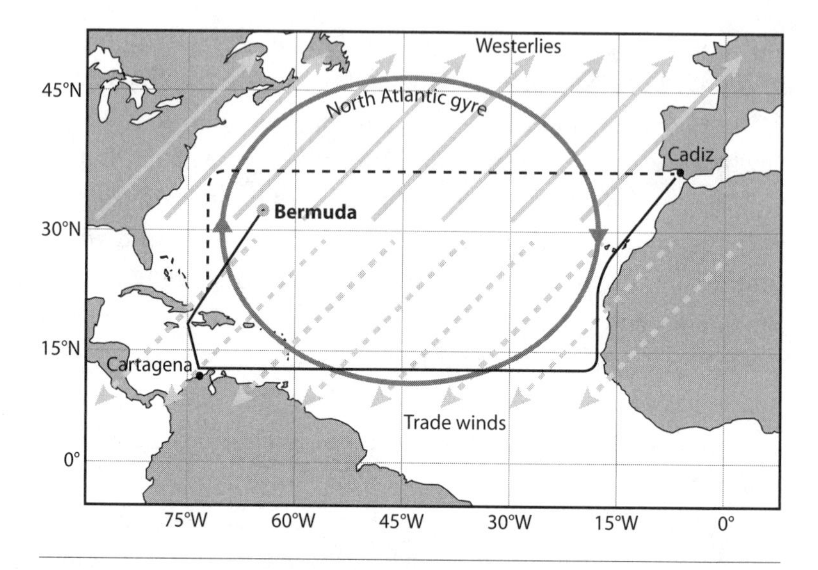

Journey of the *San Antonio*. The ship left Cadiz with a following wind and sailed south along the African coast to the latitude of Cartagena. Then it used the trade winds and the North Atlantic gyre to propel it to the Caribbean. The return voyage was to take it north to the latitude of Cadiz, then east with the westerlies and gyre at its back to Cadiz (dashed line). Unable to measure its longitude, the vessel in fact drifted east and ran aground on the reefs of Bermuda. (The rectangular projection used here exaggerates east–west distances at higher latitudes.)

their first season, before they learn enough to plot routes that incorporate the subtle realities of spherical geometry.

Going south before "westing" to the New World allowed sailors to take advantage of the trade winds, which blow generally from the northeast in the northern tropics. The resulting "broad reach," with the breeze coming diagonally from behind, extracts the maximum force from the wind. The ocean current at this latitude, part of the North Atlantic gyre, also provides a welcome two-mile-an-hour push to the west. Hundreds of species of migrating birds, sea turtles, and fish also know how to take advantage of these winds and currents.

The problem for sailors came with the return journey, their ships laden with treasure looted from Mexico and Peru. September is the height of the hurricane season in the Atlantic; the *San Antonio*, one of many vessels waiting in Cartagena, did not load her share of the booty until late August. It was a rich haul: thousands of hides, 6000 pounds of indigo, 30,000 pounds of tobacco, 5000 pounds of sarsaparilla, and 5000 English pounds worth of gold and silver. But there was not a moment to lose before bad weather set in.

Unfortunately deVerar and his convoy could not simply sail back east; that would take them into the teeth of the wind and current. Without wings or fins to propel them, they instead had to return to the gyre and sail north amid often contrary winds, piloting their way through the islands of the West Indies. The captain's plan was to use his magnetic compass in a two-step strategy of vector navigation. The compass would allow him to bear north to the latitude of Cadiz and then (resetting his course to the east) let the westerlies and the gyre carry him across the Atlantic. As the *San Antonio* left the last of the well-charted Caribbean islands behind and plunged north toward her turning point at 36.5° north latitude, she necessarily lost track of her exact position.

The convoy depended on occasional sightings of the sun and stars for latitude; for longitude they had to rely on *dead reckoning*, a common procedure used by animals that we will look at in detail presently. This all-too-appropriately named strategy keeps track of approximate headings and speeds and times (with, in the case of humans, nothing more sophisticated than a compass, a knotted rope, and an hourglass), and then attempts to reconstruct location by integrating over the various legs of the journey. As with any bird or fish in the same situation, small errors in judging distance, direction, time, or velocity inevitably accumulate, making the resulting estimate ever less accurate. And once out of sight of land there is no way to factor in the drift induced by currents or wind. A sideways drift alters the actual direction traveled; a drift along the axis of travel—equivalent to a headwind or tailwind while flying—changes the distance covered.

The part of the gyre they were in as they left the Caribbean is notoriously unreliable. They might be in the center of the flow, being carried north or northeast at 3 mph; they might be a bit to the east in the Sargasso Sea, the huge calm eye of the gyre; or they might be in the Gulf Stream, which peels off unpredictably to the NE, warming Europe as it carries tropical waters to the British Isles. To make matters worse, the convoy was racing before a tropical storm, and had gone without a sighting of the sun or stars for some days.

By the first hours of September 12 deVerar's ship was at 32.3° latitude, about 250 miles south of their intended right turn. Unaware of the branch current that was propelling the ship and with no celestial sightings (the clouds hid the first-quarter moon, which in any event had set at midnight), the navigator's dead reckoning placed them about 50 miles south of his actual position—a harmless enough mistake on its own. The rest of the convoy, scatted by the storm, was about 20 miles behind and slightly to the west. Unfortunately, the *San Antonio* and the other ships also had drifted about 100 miles east of the northerly track they were trying to maintain, carried by a warm offshoot of the Gulf Stream. The tide was at its highest about 2 a.m., just covering the treacherous reefs.

Just ahead, the "devils" had taken wing about three hours earlier. Nocturnal gadfly petrels known as cahows spend their nights flying low over the water in search of squid, punctuating their hunting with loud, eerie screeching. The lights aboard the *San Antonio* had drawn them like moths as the ship pushed blindly NNE, just to the west of the islands that were home to the seabird colony. The terror inspired by the unearthly shrieking of the birds combined with the navigational incompetence of her human pilots and the shortcomings of their instruments to doom the ship.

San Antonio's track. This satellite view shows the chain of islands that make up Bermuda and (in light gray) the surrounding reefs, which are mainly to the west and north of the islands. The ship was sailing NNE, entering this picture in the lower-left corner and striking a submerged reef about two miles west of the mainland.

Unlike the sailors, the cahows are superb navigators. They seem to know their longitude to within a mile and their latitude with even better precision, clouds or no clouds. Their internal compasses are far more reliable than anything the hand of man could produce at the time. The islands were the birds' breeding grounds; this was their one landmark in a sea empty for hundreds of miles, a beacon of safety rather than danger. And the cahows are by no means exceptional. The ocean around Bermuda is full of equally adept navigators, many of which know just where they are at any given moment. Green and loggerhead sea turtles, white-tailed tropicbirds, American eels, yellow-fin tuna, and humpback whales—none of these animal navigators are in danger of losing track of their position for long. Only our species earns this dubious distinction.

An animal's ability to know its location and the direction of its goal is one of the greatest mysteries of science. Increasingly, though much remains to be discovered and understood, this ability seems less magical. Some of the mysteries are merely products of our opposing desires to romanticize behaviors on the one hand and oversimplify them on the other, and thus to look in the wrong places for answers. We also are prone to anthropomorphize, imagining that animals see challenges in the same way we do and use the same strategies to solve the problems they encounter. As a result, we have often overlooked some surprising alternative approaches that make complex tasks much simpler for well-programmed animals. In particular, we have assumed that our fellow creatures cannot measure orientation parameters any more accurately than human instrumentation, and ought in fact to do less well than our elaborate and expensive equipment.

Our plan is to look briefly at the range of orientation strategies evident in animals, from the simple to the astonishing. Because an essential component of many of these strategies is the ability to measure periods and intervals, we will examine time sense. The next most basic component is an array of alternative compasses to orient movement. We will then look at how time and compasses combine with memory to permit piloting and inertial navigation. This will lead us to the greatest challenge to human understanding, the map sense. With a fuller picture of how animals navigate, we will conclude with the imminent threats humans pose to navigators: habitat destruction and climate change.

As we will see, understanding animal navigation is often critical to conservation, and the recent gigantic steps in decoding the workings of the compass and map sense have come not a moment too soon. Consider the plight of the cahows, whose superb navigational skill and unearthly voices proved no defense against hungry colonists or their rats, cats, and rooting hogs.

Gurnet Rock, off Bermuda. It was on a desolate islet like this, half a mile to the west, that the last colony of cahows clung to life. (The exact location of the refuge island remains a secret.)

The wholesale slaughter of the birds continued despite one of the earliest efforts at conservation, a 17th-century edict by the governor against the "spoyle and havocke of the Cohowes." They were thought to be extinct on the mainland by the late 1620s, less than 20 years after the islands were first settled.

After more than three centuries with no confirmed sightings, however, 18 nesting pairs were discovered on a small, inhospitable islet on the southern fringe of Bermuda in 1951. None of their young survived that year, decimated by rough weather, competing tropicbirds, and scavenging rats. After an intensive breeding effort there are now 250 individuals, their survival dependent on dedicated conservationists who transfer the chicks painstakingly to handmade burrows on larger islands.

Our knowledge of the cahow's behavior, and particularly the way the chicks imprint on the location of their burrows and the coordinates of their particular tiny island, is still a bit sketchy. Each pair lays only one egg. Several nights after its parents leave for the open ocean the fledged but inexperienced chick walks out of its burrow to the edge of a nearby cliff, looks at the stars overhead, makes some mental measurements (probably of magnetic field strength and inclination), spreads its wings, and plunges into the dark to take up its destiny as a wandering seabird. Without specialized instruments, charts, or tables, the cahow depends day after day on its innate ability to orient and navigate the North Atlantic. Five years later each steers its way back to Bermuda to breed. How do they do it?

■ Getting Warmer

Cahows are, to be sure, pushing the limits of evolutionary technology. For most species of navigators and migrants the challenges are less extreme, though every bit as important. Sightless coral larvae spawned on the reefs just off Bermuda must make their way up near (but not too near) the surface at the right time of day to avoid predatory reef fish, feed for a few weeks, and then return to the reefs to find a suitable place to settle and start growing. Bacteria, protozoans, and plankton that inhabit the island soil and the inshore waters also move up and down, responding to cues that indicate danger or safety, food or toxins, forever improving on their location as best they can in a changing world. The bees tirelessly carrying pollen and nectar from the island's semitropical vegetation back to their hives are among the most elegant navigators on the planet.

To combat what he saw as an anti-intellectual wave of anthropomorphism in animal psychology, the 19th-century psychologist

C. Lloyd Morgan asserted that when multiple theories compete, the explanation that introduces the fewest steps is likely to be the best. The sensible injunction to researchers to account for an animal's behavior in the simplest possible manner is known as Morgan's Canon. In the first half of the 20th century behaviorists even used this mantra to account for human behavior entirely in terms of conditioned responses to stimuli.

Because animals were seen as little more than machines, researchers invoked the canon to explain navigation and migration in terms of automatic responses to immediate environmental cues. Whether in bees or bacteria, petrels or protozoa, much of orientation and navigation is based ultimately on a limited set of sensory cues and processing tricks. But while these strategies are largely inborn, it does not follow that they are simple.

Zooplankton, the minute drifting organisms in the sea that ultimately feed nearly all of the ocean's fish, migrate down daily and back up at night. The logic of this redeployment is simple. Their prey—the photosynthetic bacteria, protists, and algae collectively known as *phytoplankton*, or drifting plants—are near the surface around the clock. But the many fish with excellent vision that eat the zooplankton are generally *diurnal*, active during the day. For copepods and other zooplankton, dining on phytoplankton at night and plunging into the relative safety of the depths during the day makes perfect sense.

The massive migration of zooplankton has been simplistically explained as an alternation between an aversion to light when the sun is available, and a balancing aversion to gravity when the sun is not present. In fact, however, the zooplankton begin their journeys *before* light levels change, actually anticipating dawn and dusk.

While much of animal (and human) behavior is in fact rooted in relatively simple responses, navigational abilities typically depend on sophisticated processing as well as multiple sensory and endogenous (internal) inputs. Even with microorganisms, for in-

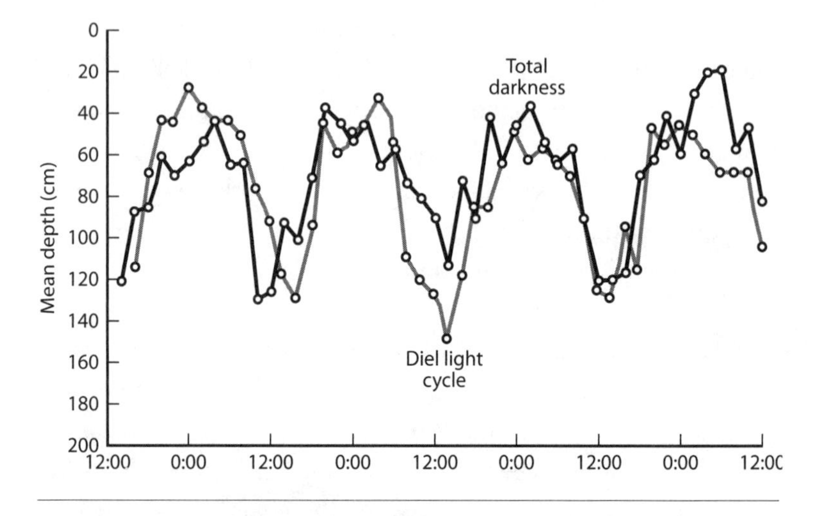

Vertical migration in zooplankton. Tiny crustaceans known as copepods migrate up to feed on phytoplankton (photosynthetic microorganisms) at night. During the day when the fish that feed on copepods are active, the crustaceans swim down to the darkness of the deep water. The copepods perform the same vertical redeployment in the laboratory in the absence of external cues. (Figure drawn from description of plankton minitowers appearing in Bochdansky and Bollens [2004].)

stance, something far more interesting than simple reactions is going on. Zooplankton moved from the ocean to a darkened aquarium still redeploy down in the daytime and up at night, at least for a few days. This surprising persistence of vertical movement is not based on some trick for sensing the sun through solid walls; instead the organisms have a 24-hour timer that has "learned" which are the daytime hours and which are associated with night. The rhythm persists in constant light or dark. The internal timer, not the appearance and disappearance of the cues, controls the creature's response preferences. Shift dawn artificially by a few hours and plankton experience jet lag, only slowly getting back into phase with the sun.

Movement directly toward a cue is called a *taxis*. *Phototaxis* (toward light) causes hatchling sea turtles on a beach to move toward the relatively bright horizon out to sea (as opposed to the darker horizon inland); negative phototaxis sends roaches scurrying away from light. Zooplankton employ negative phototaxis to orient down to safety in the daytime. A *negative geotaxis* (away from the earth) takes them back up at night. But taxis is just a word; the underlying behavior is not necessarily simple, nor consistent between creatures. Geotaxis in some bacteria, for instance, is based on magnetite grains: miniature magnets rotate the mud-loving organisms automatically and point them roughly down. Some single-celled algae have clumps of dense starch grains at one end weighing them down; other species have a low-density oily bubble that signals the way up. More complex animals—ourselves, for instance—devote an entire organ and many neurons to determining the direction of gravity. A reductionist might try to describe a human's erect stance as negatively geotactic, but we are none the wiser as a result.

Unlike the nearly universal orientation to gravity, most taxic behavior is based on comparing two or more measurements. The measurements can be made simultaneously or sequentially. Perhaps the simplest taxis known, and doubtless the oldest evolutionarily, is the positive *chemotaxis* that leads bacteria up a food gradient. The behavior is based on a sequential comparison overlaid on a biased random walk. Bacteria have to use time comparisons; they are too small to sense a difference between the concentration of a chemical at their front versus their back ends. To measure a gradient, these microorganisms must sample the sugar concentration at different places and remember whether the last measurement was better or worse than the current one. Most bacteria move by means of rotating flagella that act as propellers. Every few seconds the bacterium pauses and sets out in a nearly random new direction. They have a simple rule: if things are getting better, delay the next reorientation; if things are getting worse, try a new direction sooner.

This "getting-warmer" strategy requires at least brief memory. It is by no means restricted to microorganisms. A male moth look-

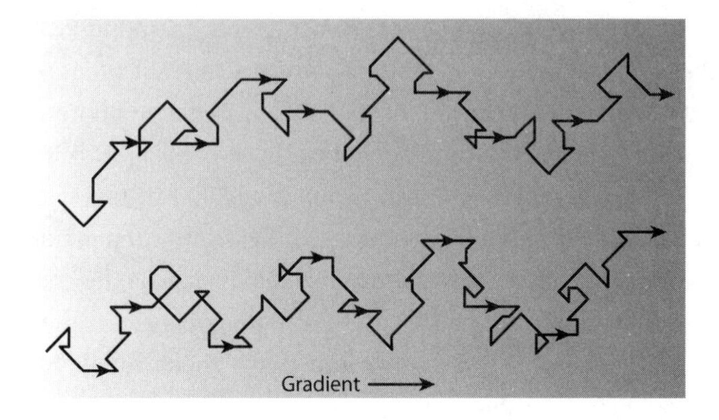

Bacterial chemotaxis. Two bacteria are shown moving up a gradient to the right. Each bacterium moves along a track but periodically stops and adopts a new direction. The length of each leg depends on whether the bacterium senses that the concentration is increasing or decreasing, and at what rate. Thus the bacterium persists in directions that take it up the gradient, but quickly abandons orientations that take it down.

ing for a mate locates the source more systematically, flying long transects across the wind searching for a trace of the sexual pheromone of his species. When his antennae sense the odor he turns upwind (positive *anemotaxis*). But even as he works his way up the scent trail he casts back and forth, finding first the right-hand edge of the odor plume, and then the left. This makes sense because the wind direction may have changed gradually (producing a curved plume) or rapidly (generating a nearly discontinuous one). As with bacteria, the moth's tracking behavior depends on successive measurements of odor concentration. We do something similar when we try to locate the source of an odor or the location of a heat-producing object.

More common to our human experience, however, is the simultaneous-comparison approach. Although low-frequency localization depends on analyzing the time delay between our ears, we localize high-frequency sound by comparing the intensity in the left ear versus that in the right. A positive *phonotaxis* (toward sound), as we observe when female crickets or frogs move toward singing males, or the negative phonotaxis of moths trying to avoid the sonar clicks of hunting bats, involves orienting so as to equalize the intensity or time of arrival of cues that are being measured simultaneously by the two organs. A marine or terrestrial flatworm displaying positive phototaxis is doing the same thing, equalizing the light intensity reaching its two simple eyes. If the light is brighter on the right the animal turns right until the left eye is equally stimulated, and thus tracks its way toward the sun.

These appealingly simple taxis-based behaviors cannot, however, explain most aspects of animal orientation. The more that is known about orientation behavior, the clearer it becomes that animals are using much more sophisticated processing to maneuver around their worlds. A sea turtle in the Atlantic can steer a steady course toward magnetic east when in a particular range of latitudes, while a truly taxic turtle would have to choose between north (positive *magnetotaxis*) and south. An ant making its way to

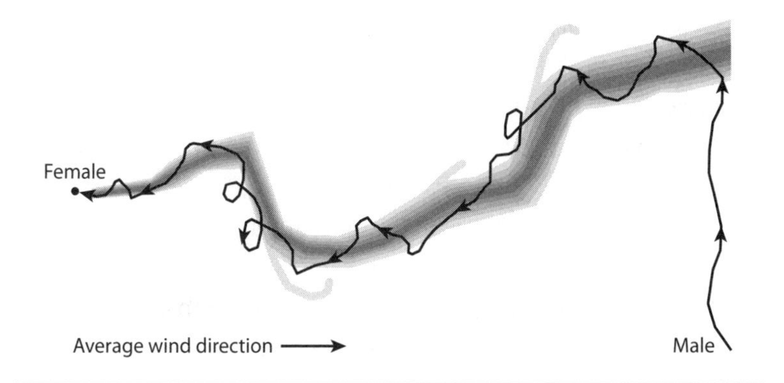

Positive anemotaxis. A female moth at the far left releases a pheromone into the breeze, which moves generally but irregularly to the right. At the far right a male moth is flying a crosswind transect until he detects the odor. He then turns into the wind, retaining some of his crosswind momentum. When he exits the odor plume he turns back and flies crosswind until he picks up the trail again, reversing course when he cannot locate the scent within a few seconds.

Table 1.1. Common Orientation and Navigation Strategies

1. Taxis	Orienting directly toward or away from a cue
2. Compass orientation	Maintaining a constant bearing relative to a cue, or a constant absolute direction if compensating for cue movement
3. Vector navigation	Using a sequence of compass bearings to steer a course; generally independent of landmarks
4. Piloting	Navigating relative to familiar landmarks; may or may not involve compasses
5. Inertial navigation	Dead reckoning; keeping track of each leg of a journey to compute location later; generally independent of landmarks
6. True navigation	Navigating with an apparent knowledge of the location of a distant goal; generally independent of landmarks

and from a familiar food source may need to keep a particular tree 135° to the left on the way out and 135° to the right while returning home. Simple taxes alone cannot account for such behavior.

Even with animals whose lives seem simple, then, there is more to navigation than taxes. And when we scale up to the challenges faced by insects, birds, whales, and humans, the need for systems of increasing sophistication to meet the demands of almost infinitely complex problems becomes inevitable. Humans have developed clumsy ad hoc strategies for navigating, and we will use these as models to understand what animals are doing with far greater ease and elegance.

The first analogy we need to call on is the compass, which humans use to maintain a constant bearing in the world. Because the main animal compass—the sun—moves through the sky, humans and animals alike need some way to measure time to move beyond simple taxes. Memory adds more power to the navigational system: sequential compass bearings linked to timing systems can provide the vectors that guide migrating songbirds. Like humans, animals too can memorize landmarks in earth, air, sea, or sky to provide piloting information. Many animals also can keep track of the distances and directions on the outward legs of a journey to compute location through dead reckoning. Then, for those creatures lucky enough to know where they are on the face of the globe, "true" navigation can take them to the actual location of a distant goal. Each layer of processing enhances the survival prospects of animals that must travel to feed or reproduce.

In the chapters to come, we will discuss each system and the creatures that employ it in an effort to understand what senses and abilities they call upon, and how they manage such incredible feats.

Chapter 2

When and Where

Taxes are powerful forces that drive much of behavior, particularly of microorganisms. But a monarch butterfly heading south toward an isolated mountain peak in Mexico a thousand miles away can do much more. It will steer to the right of the sun at 9 a.m., toward the sun at noon, well to the left of the sun at 3 p.m., and so on, always heading due south. This remarkable behavior is commonplace among animals, and is even more elaborate than it sounds. The exact direction of south relative to the sun depends not just on the time of day, but on the date and latitude as well, complications that appear to pose no great difficulty for either insects or nonhuman vertebrates. Nor does compass orientation necessarily depend on a particular cue. Animals employ a similar and equally flexible compass-bearing ability when they use stars, polarized light, magnetic fields, and visual landmarks—even when they compensate at the same time for crosswinds or currents. Our goal in this chapter is to take a preliminary look at the range of navigational and mapping strategies available to animals.

Taxis-based orientation is the simplest strategy, a special case in which the goal happens to lie directly toward or away from a cue, as is often the situation when orienting to gravity or away from danger. For the monarch, whose wintering spot may be in any of an

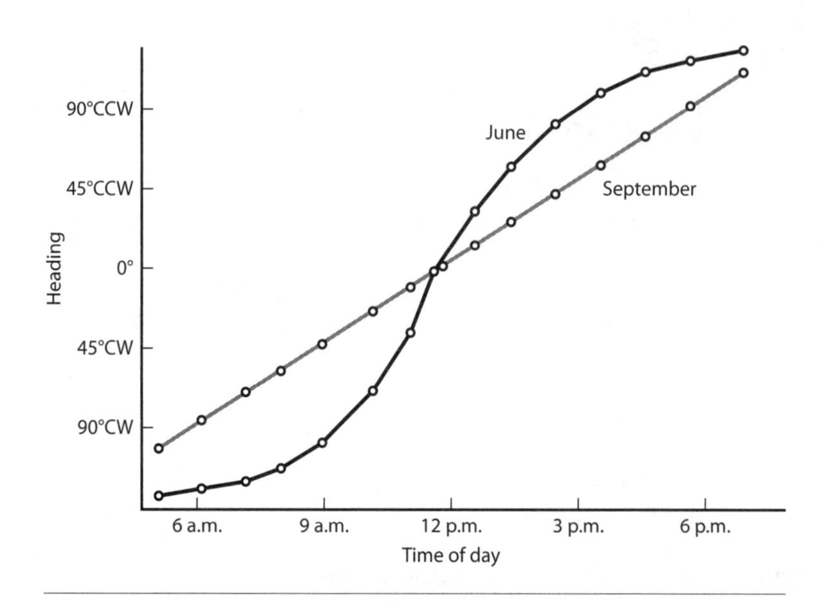

Changing compass bearings. A butterfly attempting to fly south at 45° north latitude in late June must steer to the right of the sun in the morning and to the left in the afternoon. The exact angle (plotted as a curving line of circles) depends on the time of day, and changes slowly near dawn and dusk but more rapidly near solar noon. By contrast, the same butterfly in late September must adjust for the sun's apparent movement in a different way (diagonal line).

infinity of southerly directions from its home range east of the Rockies, a much more sophisticated system is at work.

A number of long-distance migrants employ vector navigation, a more intricate strategy that links together two or more legs of their journey that may not point toward the goal at all. Several species of birds that breed in the Arctic, for example, fly first one and then a second (and in some cases even a third) fixed compass bearing south for their first winter. Red-eyed vireos, for instance, breed all across Canada, and then winter in South America. Most vireos heading south in the fall take up a SE bearing if they are coming from the west but adopt a SW vector if they are starting from eastern Canada. Along with populations from central Canada, all converge on the south-central United States before joining the "flyway" south and then SE to South America. Why they fly a dogleg is unclear, but it may serve to keep them traveling over the best areas for en route foraging.

Though simple, the strategy of using a chain of straight lines is inefficient; to minimize distance the lines should actually be gentle curves. And indeed some species learn enough on their first journey to fly a much shorter path on subsequent trips, a strategy that may require frequent changes of course if the best route is an arc.

The many species of songbirds breeding in Canada absorb enough information during that first fall journey to convert from

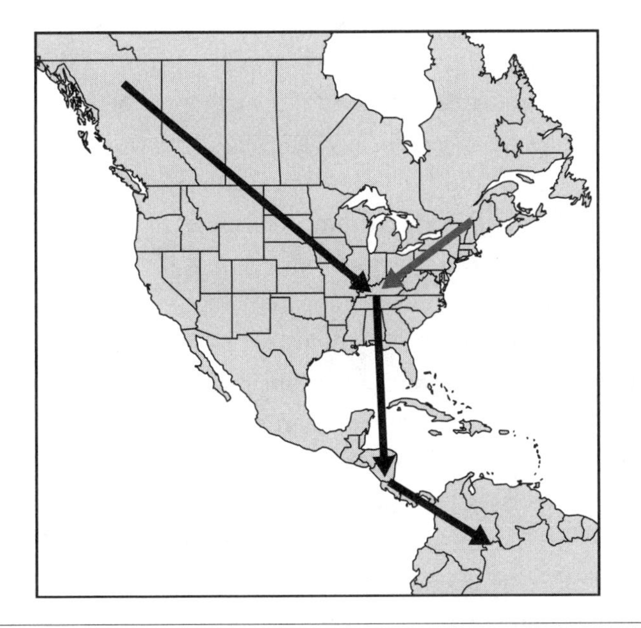

A dogleg route south. Many populations of red-eyed vireos migrating to their wintering grounds from western Canada fly SE into the United States, then south to Central America, then SE again into the Amazon. Most populations from eastern Canada set off SW, and then join the main route in the south-central United States. (The Mercator projection used here shows true compass bearings, but progressively exaggerates distances and land areas in more northerly regions.)

a set of innate plane geometry coordinates to a much more elaborate system based on a kind of spherical geometry—a daunting challenge but essential to long-distance human navigators before the invention of GPS satellites. For humans, even sailing along simple but inefficient straight lines (other than north–south and east–west) represented a breakthrough. Because the earth is a sphere, it's not possible to draw a map on a flat piece of paper without distortion unless the area covered is very small. Early maps used a rectangular projection that preserved latitude and longitude as horizontal and vertical lines but ignored the resulting distortions of shape, area, and compass directions other than due north, east, south, and west. For navigators, the first big step forward came with the discovery in 1569 by the Flemish cartographer Gerardus Mercator that all true compass bearings between any two locations could be preserved if the distances between latitude lines on a map are systematically exaggerated toward the poles. The resulting constant-bearing route is called a *rhumb line* (or, more technically, a *loxodrome*). A rhumb line appears to be a straight line on the familiar Mercator projection, but on a globe its true character as a curving track becomes apparent.

For sailors the possibility of using a Mercator map to sail a geometrically straight route on a globe may sound like a great simplification, but though it saves some compass calculations, the rhumb-line route makes plotting the ship's position more difficult, at least initially. As we will discover, open-ocean navigation depends on using estimates of direction, drift, speed, and time to estimate movement since the last good positional fix. This dead-reckoning approximation is used to plot a location on a map. But if the map uses a Mercator projection, both north–south and east–west distances depend on latitude in a complex way. In fact, converting estimated displacement into Mercator coordinates became theoretically possible only in 1594 when the first table of sines was published for sailors. In principle this allowed a navigator to calculate latitudinal and longitudinal distortions, but the computations were difficult. The conversion finally became practicable in 1604 when the Scottish mathematician John Napier discovered logarithms (the basis of the slide rule, a device developed in 1622).

Despite their relative simplicity, rhumb-line paths are almost always longer, and frequently *much* longer, than straight lines on the globe—the so-called *great-circle* routes. Great circles are the province of spherical geometry; as any student of high school mathematics will attest, spherical computations are far more challenging than the calculations of ordinary plane geometry. The easiest way to represent a great circle is with a conic projection, another common basis for drawing maps. Imagine a cone placed over the globe so that its surface contacts the earth along a circle connecting the destination and departure points. Project all lines up from the center of the earth onto the cone, and you have a map that looks more like what an observer would see from space than any other practical method.

Some birds use the Mercator strategy all their lives, but others graduate to the more efficient (but computationally more challenging) great circles. Vector navigators, whether they use rhumb lines or compute great circles, require compasses to orient every moment of the journey. They also must be able to measure or estimate distance traveled, either directly or by combining time and speed. We should keep in mind that like all the formal categories of orientation behavior, "vector navigation" is just a name. In each category there can be a broad range of complexity and underlying processing. Moreover, animals frequently fail to respect this tidy hierarchy of strategies; they may mix two together at different stages in the journey or even at the same time. When a critical cue disappears, an animal may need to switch to another, and perhaps to a different plan of attack altogether. But the most common occasion for a major strategy change involves the parts of the trip that take the creature within a familiar area.

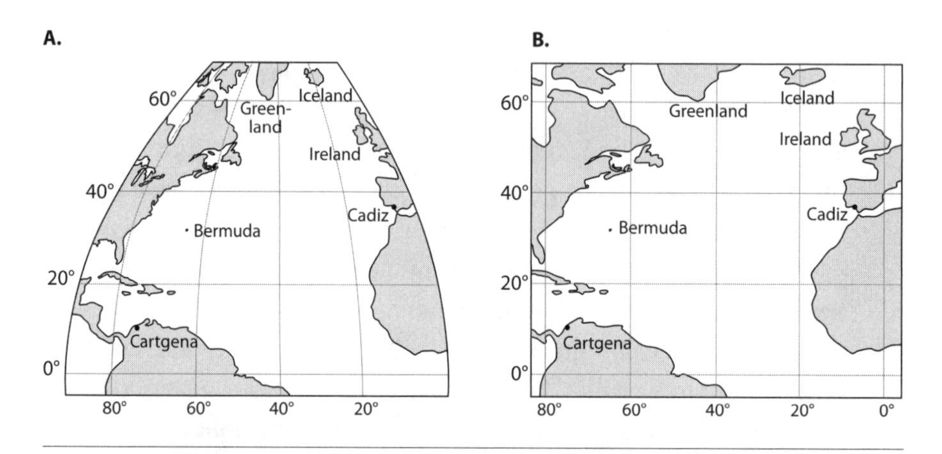

Map projections compared. If we could peel off a limited part of the earth's surface and lay it flat, the result would resemble a sinusoidal projection (A), a map that preserves areas and horizontal latitude lines. A rectangular projection (B) preserves vertical longitude lines as well as latitudes, but exaggerates horizontal distance at higher latitudes (look at Iceland and Greenland); this is the most common projection on current small-scale maps, and was widely used on global maps until the Mercator projection (C). The Mercator also preserves latitude and longitude orientations, but exaggerates both

■ Mapping and Map Use

Serious sailors often distinguish between coastal navigation and the more challenging blue-water sailing, which takes them out of sight of land. Vector navigation is one of several blue-water strategies. The first human navigators at sea, on the other hand, were necessarily experts in the art of coastal navigation—more generally known in orientation research as local piloting (or *pilotage*). A ship's captain, for instance, steers his boat past a tricky headland using visual landmarks ashore to judge his position in the water. For casual sailing in familiar waters the process is based on mem-

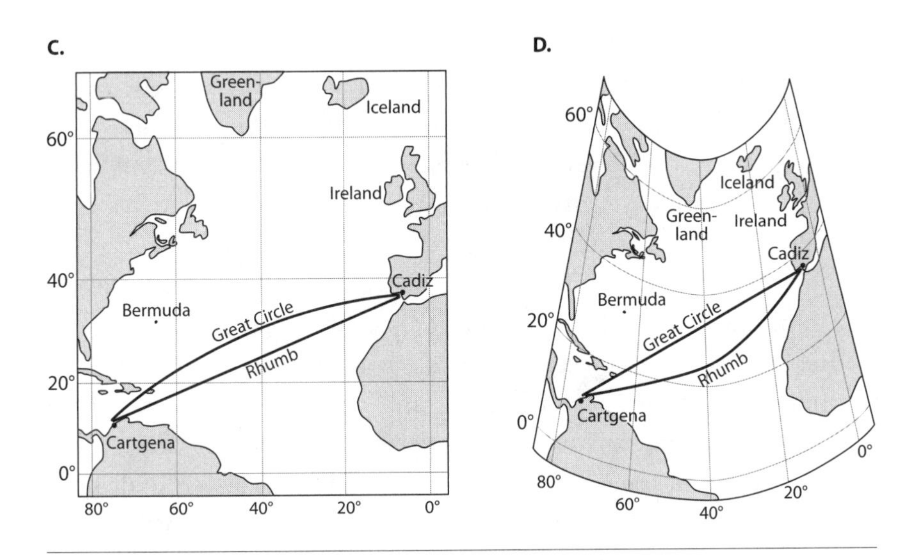

horizontal and vertical distance at higher latitudes (again, look at Iceland and Greenland). The genius of the projection is that a constant-bearing route generates a straight line on the map. The rhumb-line route (247.5°) from Cadiz to Cartagena is marked, along with the shorter great-circle route, which appears as an arc. The most common representation now for moderately large areas is the conic projection (D), which when properly centered shows the great-circle track as the straight line that it is, roughly as it might be seen from space; the less efficient curving rhumb-line route is also shown.

ory. Most of us pilot our way back to our car in a parking lot more or less expertly, or from one store in the mall to another in just this fashion, barely thinking about what we are doing. In mammals at least, a specialized part of the brain is devoted to this frequent and critical task. But for more complex piloting where there is less room for error or the surroundings are less familiar, the navigator may need a detailed map. He must labor with compass, straightedge, and calculator to accomplish what our built-in neural module does unconsciously in simpler situations.

For a sailor piloting a boat along the coast the process involves a combination of landmarks and triangulation. From any point in the water there is a specific angle between two prominent marks on shore. You might suppose that this solves the problem, but the same angle also can be observed from other places in the water. In the diagram two towers are separated by 24° at both locations A and A′ (and along a curving line of intermediate positions that are not shown). But with a third angle there is only a single spot where this viewpoint exists. The final reference can be a landmark (the third tower in the diagram) or a compass direction such as magnetic north. The process then is to triangulate a position, determine the location, plot a heading, and move along the heading. Another bearing fix (point B in the diagram) establishes the true path (the "run made good"), with any complicating leeway or drift becoming clear. In flying, for example, the discrepancy between where you are and where you intended to be can be a consequence of sideways displacement by a crosswind, as well as a headwind or tailwind (which will affect the total distance traveled).

The boat being piloted along the shore illustrates the three main sorts of errors animals must contend with. The first is *leeway*. The ship has an intended heading (the direction the navigator wishes to go) and an actual heading (the direction the bow is pointed). These are different because the wind that is powering the ship also pushes it to the east. The boat must head into the wind to allow for this displacement. The discrepancy (the leeway) is obvious: the sailor need only compare the axis of the ship to the wake left behind in the water.

What the wake does not reveal is whether the water itself is moving. In this case, there is a current from the southwest carrying the boat and its wake to the NE. This second error is what we will call *drift*; it's what doomed the *San Antonio*. (Sailors distinguish between the direction or "set" of a current and its speed or "drift"; for our purposes we will lump the two together.) The final problem is with the actual measurements of the three parameters needed to calculate the course until there is a new fix using the landmarks—the precise compass direction, speed, and time. (The

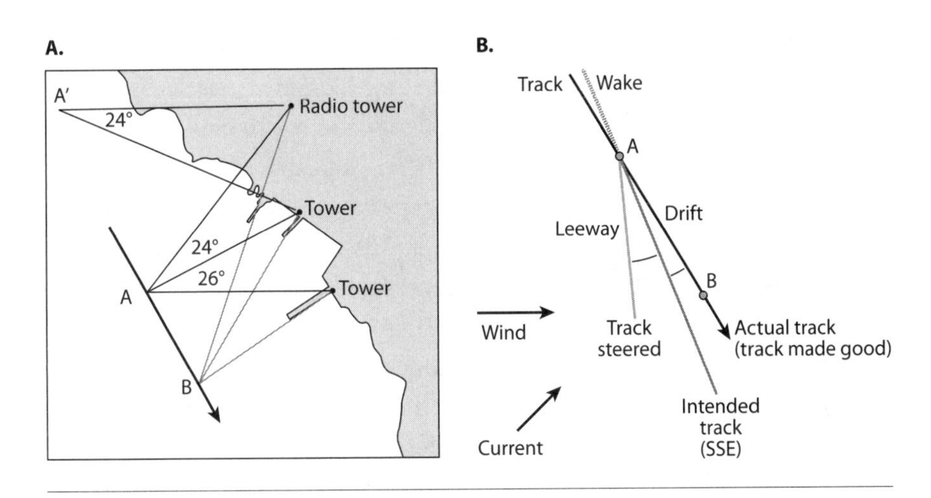

A track along a shore. (A) The boat at position A determines its location by triangulating three landmarks (one of which could be magnetic north instead of the third tower used in this example), and sets a course (say SSE—157.5°). (B) By triangulating again later the sailor determines his subsequent position, and discovers a slight northeasterly drift—an actual track of 148° in this instance, which is particularly noteworthy because it is pushing the ship toward the shore. The actual track steered is somewhat into the wind to compensate for the easterly displacement of the boat by the wind (the leeway).

navigator must combine estimates of speed and time to compute distance traveled.) Uncertainty in these measures limits the accuracy of any position finding for both humans and animals.

For convenience, we will use the same terms from sailing when we discuss animal navigation: leeway is the directional error you know about; drift is the discrepancy you cannot detect. If we apply this leeway–drift distinction to birds flying under clear daylight conditions, the leeway is the angle the animal needs to turn into the wind in order to maintain its intended track. There is no drift because the ground visible below is not shifting its position.

But at night, in clouds, or over the water, the same crosswind can become instead a source of drift across the unseen earth below. A honey bee flying out from land over calm water loses its ability to compensate for wind speed and direction; at the resolution of its eyes the texture of the ground needed for judging its track disappears. Even in daylight over a choppy sea cahows can only roughly estimate the crosswinds because the ocean current and slowly moving waves shift the visible landscape below; they can compensate for some of this relative motion in the manner of leeway, but the rest is drift. At night, all motion seems like drift.

The advantage of piloting is that an animal obtains frequent updates of its location from the landmarks around it. Leeway and drift are automatically corrected with each new fix, and thus there is no accumulation of error. The piloting process for sailors is, at least in theory, identical to what the brains of animals must do in familiar areas. Honey bees flying to and from food come to rely on intermediate landmarks, in some cases going out of their way to pass by visual signposts. Foraging wrens do the same within their territories, checking the relative locations of prominent features. The chart used by the sailor becomes for animals a mental representation called a *cognitive map*. Kidnap a wren or a bee and release it elsewhere on its home grounds and it will generally triangulate its unexpected position using two landmarks and the sun (or three landmarks under overcast) and set course home.

But birds, bees, and people also store mental snapshots—a sequence of scenes along familiar routes. If released on a well-known path an animal will often orient to this trail of pictures and dispense with the more demanding process of triangulation. This is the normal way we operate when driving near home, depending on recognizing the scene that reminds us where to turn right, where to bear left, and so on. A wrong turn throws us back on mental maps and triangulation.

We tend to think of mental maps used in piloting as local in extent, valid within range of a collection of even more local landmarks. In fact there is no reason the maps cannot extend over long distances. Migrating animals can cover thousands of miles, and if they are committing even a small subset of what they encounter to memory, their internal maps may be vast. Another point to keep in mind is that while for us landmarks are generally visual and near, other species may be able to employ different senses to detect other sorts of beacons. One possibility, for instance, is subsonic sound too low for us to hear. Pigeons and elephants, to name just two species, are equipped with ears that can pick up frequencies far below our lowest limits. Subsonic sounds can travel hundreds of miles in the air and farther in the water. In theory the noises made by winds passing over the Rocky Mountains could serve as an auditory landmark for a bird on the Mississippi. Even in the visual realm we should not take our personal sensory experience as the standard. There are animals that can see in the ultraviolet range, others that can use thermoreceptors to form images in infrared light, and yet more that can see by starlight.

Another common navigational strategy we will encounter is *inertial navigation*. We've already run across it in the truncated voyage of the *San Antonio* where it had another name: dead reckoning. Most creatures find themselves at least occasionally without familiar landmarks from which to work out their position. They might be traveling through unexplored territory just outside the home range, or well beyond their experience during the course of migration, or they could even be near home but dealing with darkness or fog. Inertial navigation requires an animal to keep track of its movements—north for two hours at 5 mph, two hours NE at 7 mph, an hour north at 10 mph, an hour east at 10 mph, then two hours more to the east at 5 mph, for instance.

In early ships on the open ocean these measurements would be recorded at hourly intervals on each watch by pegs placed on a traverse board. By adding all the legs together geometrically at the end of a watch, the navigator could calculate that the starting point in our example was about 42 miles away to the southwest. But even ignoring errors in measuring compass bearings, speed, and time, if

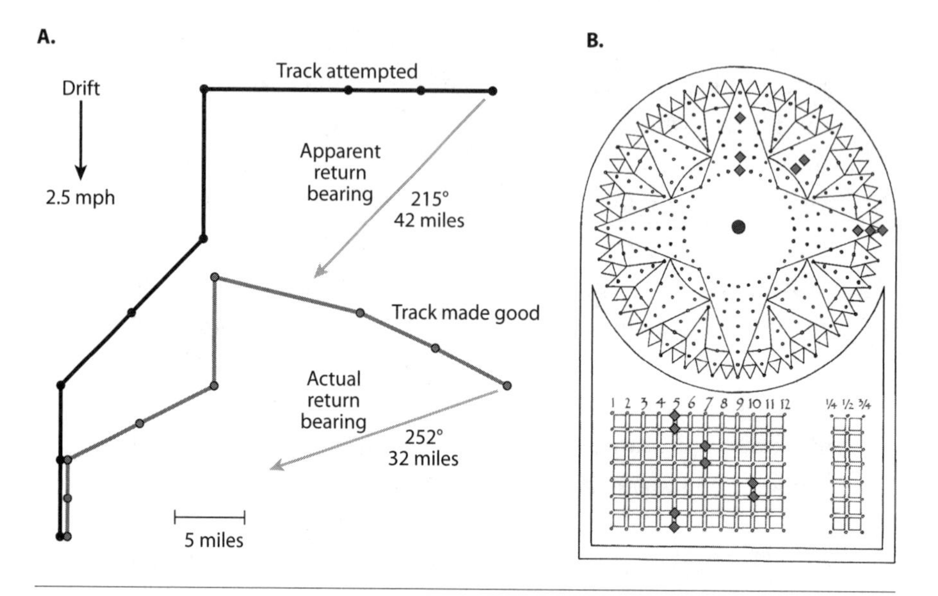

Inertial navigation. (A) In this example a sailor or animal sets off on a circuitous path involving a mix of three directions and three speeds. The navigator must then determine the net displacement by integrating the apparent legs of the journey (dark track) and then computing the homeward bearing and distance: 215° and 42 miles in the absence of errors. However there is an undetected southward drift of 2.5 mph that gives rise to an actual route (gray line), for which the correct homeward bearing and distance are quite different: 252° and 32 miles. (B) Illiterate sailors recorded bearing and speed at hourly intervals on a traverse board. An officer then transferred the numbers to a log from which the navigator calculated a position from dead reckoning.

there is drift the dead reckoning can be quite wrong. In the example shown here the drift error alone is a sobering 20 miles out of the 42 miles attempted.

To use inertial navigation an animal must have a compass so that it can record the orientation of each leg, some way of estimating distance traveled along each segment of its route, and a clock or interval timer. Early ships measured apparent speed by heaving a log overboard and counting the number of knots on an attached line that were carried out in 30 or 60 seconds. The knots were about

50 feet apart; each knot corresponded to one knot of speed—one nautical mile per hour.* Animals, of course, do not toss logs; instead bees and birds have evolved special circuits in their eyes to judge the rate at which the terrain below is moving by. An accurate calculation depends on knowing the distance to the ground—no mean feat for a flying animal.

Drift is a universal problem, but for terrestrial creatures at least there is no real effect of crosswinds or currents. Bees can use another set of special circuits in their visual processing areas to compare the observed movement of the ground below them with the heading they are attempting to fly. This allows them to judge their direction to take into account what is basically leeway. They adjust their estimate of distance to take into account the effect of headwinds and tailwinds; within limits they do this by adjusting their airspeed to produce a constant ground speed of 13 knots (15 mph). Once they are ready to return they use this data to perform a path-integration calculation. Bees have to be good at this because their poor eyesight makes most landmarks useless, at least at a distance. Sharp-eyed birds, on the other hand, can rely more on piloting and less on dead reckoning in their travels about their home range.

■ "True" Navigation

The most challenging problem for a navigator is to travel to a relatively precise target at a considerable distance without familiar landmarks for piloting. In most cases this so-called *true navigation*

*A nautical mile is 1/60th of a degree of latitude, now set at 6076 feet. Originally statute and nautical miles were the same—5000 feet—based on a first-century-BCE underestimate of the circumference of the earth. The first correction took the figure to 5280 feet, the familiar statute mile; subsequent emendations have only corrected nautical distances. Nautical speed and airspeed are frequently expressed in knots. The meter, by contrast, is approximately 1/40,000,000th of the circumference of earth measured through the poles and Paris, a value with no simple or logical meaning in terms of latitudinal angle.

requires the animal to know more or less exactly where it is as it travels—more precisely than path integration will permit. In fact, inertial reckoning from the start point cannot be involved: some animals can be captured and flown in sensory isolation well to the right or left of their route, and yet will reorient immediately to take the unplanned displacement into account. The most famous example of this is the ability of experienced homing pigeons to return directly to their loft after being carried hundreds of miles away in a new direction. A true navigator like the pigeon does not need to worry about leeway or drift, at least not on a large scale, because the animal can simply take a new global map fix. Errors, whatever the source, just do not accumulate. The bird need not be concerned about the exact distance traversed because its map sense will say when it has arrived.

It is important to keep in mind that the cognitive or home-range map is quite different from the *map sense*. A cognitive map depends on memory, on having already been at or near a spot and recalling the nearby landmarks. It requires a compass and triangulation. The map sense is more like a GPS readout, a set of parameters specifying a point on the surface of the earth that the animal may never have seen. In some way, the map system in the animal "knows" where it is, at least relative to home.

But while humans think of location in terms of north–south and east–west coordinates—latitude and longitude—there is no real reason to suppose that evolution has taken the same approach in all animals. There could be two meaningful parameters, or three or four; the map-sense coordinates of different animals might intersect at right angles like latitude and longitude or at some other angle. Indeed, an animal's goal may possess unique cues; areas at a distance in multiple directions might provide orientation signals of their own. If a cahow recognized the unique aromatic bouquet of Bermuda, for instance, then any time it was downwind of the island it would know the direction of home and could return much as a moth tracks down a mate.

We are guilty of a condescending anthropomorphism, reading into other orders of beings our own blindnesses and computational limitations. Selection seems to have led to a variety of solutions to the challenge of finding home or other goals, recruiting sensory systems unknown to science before the navigational behavior of bees and pigeons led us to the discovery of such systems. We have said about all we need to about phototaxis and other simple orientation behaviors. We turn now to animal clocks, compasses, and maps—the onboard hardware and software necessary for all other navigational strategies.

Chapter 3

A Matter of Time

■ Animals and Time

Honey bees evolved in the tropics and spread throughout Africa, Asia, and Europe; Columbus brought the first honey bees to the Americas. Honey was for millennia the only sweetener that could be kept immune from spoilage, so it is no surprise that colonies of people brought colonies of bees to their new homes. Their wax provided candles, and their well-ordered hives were models for society. No wonder humans have spent so much time perfecting our understanding (and mythology) of this one kind of insect. And the payoff has been enormous; honey bees have yielded up secrets far, far beyond the wildest imagination of those who have studied them. The most remarkable of all, perhaps, is their gift to students of animal navigation: their willingness to draw miniature maps of where they have been.

Honey bees are unique among temperate zone bees in being perennial—that is, they overwinter as fully staffed colonies of perhaps 20,000 workers plus a queen that they protect and keep warm. They slowly consume the honey stores they made from nectar and set aside for the winter, shivering collectively in the hive to generate warmth. During the spring and summer foraging bees travel as much as five miles on each trip, collecting nectar and pollen that they bring back to the hive, fanning tirelessly with their wings to condense the nectar into honey. They start early and work late, anticipating dawn from within the darkness of the hive.

The workers return accurately to the home colony in spite of being legally blind. Within seconds of departure the home tree behind is a blur, indistinguishable from the many others on the forest edge. Likewise, the field ahead is a landscape of smudges. As it flies the forager detects the movement of the grass below; above there is a bright area in the sky with a bull's-eye pattern around it. The bee is born able to recognize flowers, but must test each candidate to see if it is in fact a blossom, and happens to be supplying food at this time of day. If so, the best-understood learning sequence in nature unfolds as the forager commits to memory the relevant characteristics of the particular flower and the nearby landmarks that will allow it to triangulate this food source during return visits.

Each bee may make more than 500 round-trips daily for perhaps three weeks before simply wearing out. In this, and in plotting the maps by which they share information about the location of food, bees demonstrate navigational abilities far beyond our own. And interwoven throughout all their remarkable feats is an awareness of time. Bees must know when to prepare for winter, and then for spring. To use the sun as a compass when foraging they must know the time of day to compensate for its movement through the sky. To calculate the bearing home after an irregular flight out they need to know the distance flown on each leg, a task that ought to require knowing the duration of each part of the journey.

So it is with us. Despite every navigator's preoccupation with distances and angles, latitudes and longitudes, and headings and compasses, nearly all human navigation rests on a basis of understanding and measuring time. We need to know when to start, what direction to choose relative to the sun or stars, how long we've been moving if we wish to compute distance traveled, when to stop, or that most challenging task of all, how to determine relative time so we can deduce longitude.

At first glance these tasks seem to require two separate systems—a clock or set of clocks for reading the current time, and a timer for judging elapsed time. The vertical migration of zooplankton, for instance, would seem to require an internal daily clock to allow the organisms to anticipate dawn and dusk. Dead reckoning, at least as practiced by humans, requires instead an interval timer to measure elapsed time on each leg of a journey with some precision.

Many creatures require multiple timing systems to navigate. The epic migrations of monarch butterflies demand both a clock and a calendar. These frail ephemeral creatures fly thousands of miles from their summering grounds in the northern United States where food is abundant to a tiny remote spot in Mexico where they spend the winter clustered together for warmth until spring tells them to journey north again. The butterflies that begin the northward journey do not make it all the way. They stop to feed and lay eggs partway there, and the next generation carries on. It may take as many as four generations to cover the distance back to home territory.

The monarchs must have some sort of calendar to tell them when to leave Mexico and head NNE in the spring, and then when to depart SSW in the autumn. The week of departure in the fall differs depending on the latitude, so there must be well-regulated flexibility built into the system. They fly toward their goal, modifying their straight course as necessary to accommodate the wind and their current location. To accomplish this they need either to remember the correct angle to fly with respect to the sun at each time of day, a clock-based approach, or they must use a kind of spherical geometry to calculate the movement of the sun through the sky over measured intervals, a timer-centered tactic. Either strategy would work, and there is some evidence that animals can do both, and opt for the one that gives the best results under current conditions. Both calendar and clock need frequent resetting to overcome drift and to allow for the changing length of the day and time of dawn, values that vary with the seasons and latitude.

Clock- and timer-linked behaviors are important for activities keyed to other cycles as well. For instance, there are creatures for which an ability to time tidal or lunar cycles is critical. Aquatic Bermuda fireworms, for instance, emerge synchronously from their burrows in the mud near shore to mate in a brief burst of bioluminescence in the shallow water once each lunar month during the summer. The consensual moment is 57 minutes after sunset on the third evening after the full moon, and the fireworms are eerily consistent. For this, we would need a 27.3-day lunar clock to determine the date, a 24-hour daily clock to judge when the sun goes below the horizon in case the scheduled day is cloudy, and an interval timer to measure the 57 minutes from the observed or inferred sunset. A number of marine organisms also display 12.5- and 25-hour tidal cycles. While we will dissect maps and compasses in considerable detail in later chapters, we look more closely here first at the timing mechanisms (and the time-independent work-arounds) upon which most navigation and migration relies, and then at the ways in which the clocks and timers are calibrated and synchronized to overcome temporal drift.

What evidence is there that animals can use multiple timers? Humans generally talk about time both in terms of repeating periods and as unique intervals between events—quitting time, say, versus how long it has been since that last coffee break. We naturally assume animals must experience the world in the same sort of multitimer way. The most efficient strategy for a human would be to have separate timers, one for intervals and others for each kind of repeating period—a stopwatch, plus individual clocks for the week of the year, day of the lunar month, the hour of the tidal cycle, and the time of day. But few of us in fact carry around stopwatches or tidal clocks; we try to get by with a wristwatch or wall clock. Perhaps animals do something similar, using one timepiece and making it work for all applications. Is it possible that period and interval behaviors are, in fact, controlled by a single timer? If so, is it a clock or a stopwatch?

Intervals could, at least in theory, be measured by taking two successive readings from a single clocklike period timer, much as we use watches to time events when a specialized stopwatch is not at hand. Or, in principle, interval timers could serve all purposes by adding up small bits of time. If an unconscious counter kept track of the accumulated number of minute intervals, a person could use units of 60 to estimate the number of hours since dawn, employ sets of 1440 to judge days, and use larger groupings to count the number of days since the winter solstice. Most human clocks, in fact, count 32,768 vibrations of a quartz tuning fork and then advance the second hand one step; minutes are judged by adding seconds, and so on. Our wristwatches are just elaborate interval timers without the stopwatch readout.

The first question, then, is whether navigating animals are using two or more different kinds of timers, just one kind in two different ways, or have both options available and choose one or the other based on circumstances. While period- and interval-based behaviors could simply reflect two strategies for using the same basic circadian clock, opinion has shifted toward the possibility that computing periods and intervals might depend upon at least two separate circuits and basic processes, and perhaps more. Some researchers even argue, and we agree, that the four common period measurements of the animal world—annual, lunar, tidal, and circadian—may rely on independently evolved mechanisms. The yearly clock that tells birds when to migrate may have no relation to the daily one that tells the bee in the darkness of the hive that dawn is near, or the bat in its cave that dusk is coming. The system that detects the lunar cycle that drives fireworm mating and the tidal clock that sends starfish back into the waves to avoid stranding may each have evolved separately. Even this may be an underestimate: monarchs have a minimum of two separate circadian timers, one for general physiological needs such as feeding, and another for navigation.

At present, the best way to learn about these systems is a close study of the way animals sense and use time. For instance, all timers are inaccurate; they drift in and out of phase. But period clocks should accumulate errors from the last resetting, whereas interval timers should build up inaccuracy only from the beginning of the current interval. If we devise clever tests we can hope to distinguish which time-based navigational strategy is at work, and determine its temporal precision.

Periodic Clocks

The timers that affect an animal's behavior, regardless of their cycle intervals, have a few features in common. They are *endogenous*—that is, they are internally regulated—with a default rhythm that only approximates the actual cycle. The *circadian* ("about a day") timer of an average human, for instance, consistently judges the day to be 23–25 hours long. A prisoner in a lightless dungeon or a subject in a deprivation test wakes spontaneously about every 24 hours, but his personal dawn will drift systematically early or late. This drift usually determines whether an individual is a morning person whose internal rhythm is shorter than 24 hours, so that she typically wakes early, or a night owl whose endogenous timer runs slow.

Because of the drift of our free-running rhythm, we need to recalibrate frequently using external cues. Older readers will remember the morning ritual of winding and resetting wristwatches before the advent of battery-driven quartz mechanisms. An animal's capacity for daily recalibration—*entrainment* as it is called—is the other major feature common to period timers. We experience jet lag because our internal clock is out of phase with local

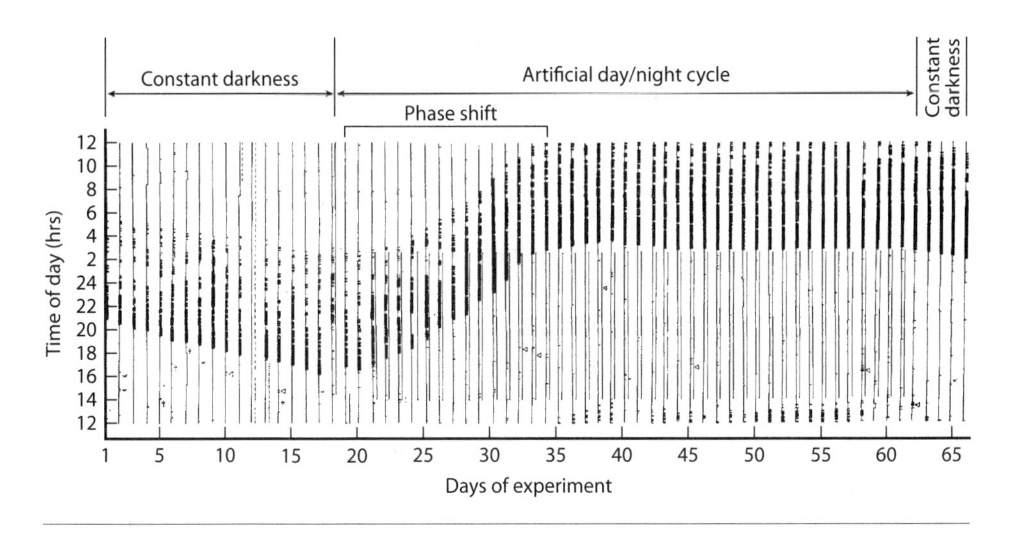

Circadian rhythm. The relative activity of a nocturnal flying squirrel held in isolation is indicated by the intensity of the vertical lines: a light band indicates rest, dark reflects wheel running. During constant darkness the animal's daily pattern of activity drifts slowly out of phase with respect to actual time; in this example the animal is starting its movement about 20 minutes earlier each day. On day 18 an artificial light–dark cycle begins. It takes the squirrel about two weeks to fully entrain ("phase shift") to it, after which the animal remains in near-perfect synchrony.

time at our destination, but slowly adjust as our bodies entrain themselves. Animals are equally sensitive to these dramatic clock shifts. Transport a hive of bees from Paris where they have been taught to collect food for a certain two-hour period—beginning, say, at noon—to New York, and they will turn up at the feeder five hours earlier local time (visiting from 7 to 9 a.m.). Each day they will shift toward true solar time so that in a week they will be feeding again starting at noon. Now that they are over their jet lag fly them back to Paris, and they will appear to be five hours late, foraging from 5 to 7 p.m. local time but shifting earlier each day.

Daily rhythms allow animals to anticipate times that are important—to start looking for food before they are hungry, for instance. Because they are endogenous the systems function smoothly even after a few days of overcast have prevented readjustment. At the extreme long end of the range we are interested in, hibernating animals such as bats, ground squirrels, and hedgehogs awaken even in the absence of external cues signaling time of year or changes in the weather. Caged migratory birds in sensory isolation become increasingly restless at night during the appropriate weeks in spring and then again in fall, and attempt to escape from the side of their enclosure that faces their migratory destination. Just as with circadian clocks, endogenous annual timers with their free-running periods prevent animals from being fooled by anomalous conditions—in this context, unusually warm or cool spring weather, or long periods of overcast.

In the wild, birds' endogenous annual rhythms are generally calibrated by the changing duration of the day or the corresponding alteration in night length. For the majority of animals dawn or dusk, depending on the species, is the most potent recalibration stimulus for circadian clocks. A major concern in the study of period timers and circadian rhythms is that researchers tend to simulate dawn by turning on the lights abruptly, and dusk by turning them off. In nature the change in light level is gradual, and accompanied by corresponding changes in sky color and polarization. The attitude of scientists is that they are making things easier for their subjects, eliminating any ambiguity about sunrise and sunset.

There is good reason to think that animals have a different perspective, though. Once they discover by chance that they can control the illumination themselves, for instance, white-footed mice in an experimental situation will invest considerable time and energy pressing levers thousands of times to make the change in light level slow and natural. Perhaps the unnatural split-second switch from bright to pitch dark interferes with the daily recalibration, and introduces an unexpected inaccuracy or uncertainty; whatever it does, it seems to be something animals would prefer to avoid.

In addition to light cues, some species can also use daily changes in temperature for calibration. Honey bees living in the darkness of a hive can even cue off minute daily variations in magnetic field strength produced as the earth warms then cools, displacing the ion-rich jet stream north and south. For modern humans and some other group-living animals, the reset button depends primarily on social factors.

■ Accuracy, Real and Possible

As we've said, most ongoing navigational tasks, at least as performed by humans, involve either knowing the time of day or measuring specific intervals, such as the number of minutes one leg of a journey has consumed. If the neural apparatus underlying both period timing (such as circadian-rhythm clocks) on the one hand and elapsed-time measurements on the other is the same, or shares major molecular components, then the accuracy of this dual-function clock must limit the precision of any and all sorts of time-dependent navigation. This would be a fortunate situation for researchers, because period timing is generally easier to study and quantify. Of course the precision of period timers is crucial to know even if animals also have an independent stopwatch with its own intrinsic errors.

In either case the inherent error (the drift) accumulates from the start of the period (the reset time) or the interval and increases inexorably. But there is a special problem with period clocks: accuracy. A 10-minute error in judging dawn, for instance, adds to the temporal drift. For animals living on the ground in anything but the most open and cloudless habitats, determining when exactly the day begins is bound to involve guesswork. The same problem arises with annual, lunar, and tidal clocks. We will look at this calibration and synchronization challenge later in the chapter.

What sort of accuracy do period timers display, and what are the consequences for navigation? Minimizing and overcoming the errors inherent in their timekeeping systems is absolutely critical to the survival of many animals. Because the sun's azimuth (its angular direction from north) changes constantly, a 15-minute uncertainty near noon on the summer solstice at 40° latitude translates into a 10° directional error in orientation. A 30-minute ambiguity means an animal could be 20° off.

How serious is this? If you were a honey bee on a typical foraging trip without landmarks to help, it could be fatal. Let's imagine your colony lives in a tree along a forest edge, the usual situation for feral bees. Suppose a patch of flowers is 500 yards away; the one-way flying time would be a bit over a minute. Let's say you visit 50 flowers in the patch in order to collect a full load of nectar or pollen, devoting 15 seconds to each blossom. That makes the total trip time before returning about 14 minutes. To minimize uncertainty, let's assume you are using an interval timer, so we do not need to worry about reset error and accumulated drift since recalibration took place. Even a 10% error in judging elapsed time (84 seconds) puts you 20 feet away from your goal, searching an unfamiliar tree in vain for the inconspicuous opening to your home cavity. With 20/2000 vision, one oak will look very like another. For a bird, make the foraging distance 10 miles and, allowing for the higher flight speed and better visual resolution, the situation is essentially the same: 10% is not nearly good enough.

For humans the typical circadian error is a devastating 60 minutes after 24 hours. If this is our basis for judging time of day, we are fortunate not to be regular celestial navigators. A nocturnal flying squirrel's 20-minute ambiguity is less extreme, but is still a serious potential source of error.

Unless these estimates of the accuracy of period timers are misleading there must be a more accurate onboard interval timer, or some way of mitigating these potential mistakes. As we will see, however, estimates of interval accuracy are no better; indeed, they seem worse. (At least the error does not increase as the day wears on; clock drift, in contrast, accumulates steadily from the time of resetting.) Lab tests with rats and mice regularly turn up time-judgment errors of 15% or more—an impossible value that says more about the inadequacies of the tests than any navigational shortcomings of the animals.

It's possible that we are looking at the wrong behavioral measures; animals might know the time within a few minutes but just not obsess on beginning their day exactly on schedule. On the other hand, creatures might have poor-quality clock or timer hardware but compensate with clever software, or use some time-independent alternative. Or perhaps all the force natural selection can bring to bear is simply unequal to the task.

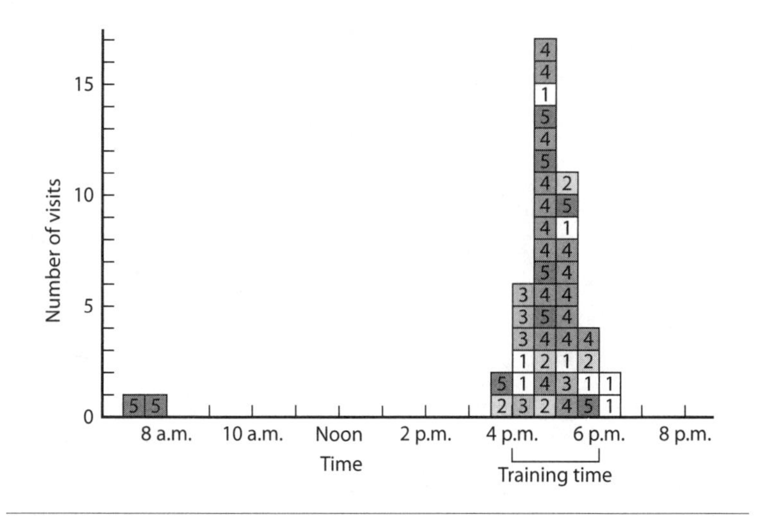

Circadian timing by honey bees. A hive of bees was kept in a basement isolated from any sight of the sun or sky. Lights went on and off on a daily cycle. Five marked foragers were trained for several days to a food source that offered sugar solution from 4 to 6 p.m. On the next day no food was offered; visits to the feeder were monitored. Bee 5 checked the feeder twice in the morning. The other 42 visits by the various bees were made exclusively between 3:30 and 6:30 p.m.

■ Intervals and Accuracy

There are strong hints that distinct periodic and interval timers are involved in many species' navigation behavior. One bit of evidence comes from the observation that the main gene that is known to be involved in daily rhythms (aptly known as CLOCK) in creatures ranging from fruit flies to mice can be deleted without affecting the ability to judge intervals. Interval timing could be essential to the sun-compensation behavior of solar navigators. Rather than using a circadian clock to judge time of day, animals could be using the interval since the last solar fix. For any particular systematic error in judging time, the shorter the interval the better. Honey bees seem to measure intervals in at least some contexts, preferring comparisons of current solar position with those made about 40 minutes earlier. More importantly, the measurement of elapsed time seems at first glance to be absolutely essential to dead reckoning: speed over a known time interval translates to map distance. But again, dissecting out the temporal precision of timers is as tricky as it was with 24-hour clocks.

The most precise—and tedious—measurements ever devised to investigate timing accuracy in navigational specialists focused on honey bees. The tests measured the flight angles of foragers departing for a familiar target after having been kept an hour without any view of the sun. During their confinement the sun moved 14–22° in the sky, depending on the date and time of day of the test. The average error of the hundreds of bees was just 0.99°. This figure is compounded of small misjudgments in elapsed time (the parameter in which we are interested), the remembered or computed rate of the sun's azimuth movement during that particular hour of the day, each bee's ability to steer by the sun using its low resolution compound eyes, and the experimenters' capacity to measure the foragers' departure bearings. Based on this data and what we know about the magnitude of the other uncertainties, our best estimate for interval accuracy in bees is about 2%.

But the fact is that we have no very good quantitative grasp on this critical parameter other than for honey bees. We also are largely in the dark about how animals contrive to outperform our lab-based expectations. We must remain alert for signs of alternative cues, covert updating, or processing work-arounds that may be available. A true map sense, for instance, could help on a large scale; timer errors would disappear with each spot check of absolute location. More locally, an ability to measure distance directly rather than by computing it from speed and time could improve accuracy dramatically.

■ Fooling the Experts

One possibility that might account for the apparent discrepancy between timer accuracy and navigational performance is that some animals in some situations just do not use time to compute distance. We are so inured to thinking in terms of the nautical solution (speed × time = distance) that we tend to overlook alternatives. "Obviously," wrote the respected expert Talbot Waterman in 1989, "in dead reckoning and vector integration such stopwatch information is crucial." The most widely used navigational text in the 20th century (*Dutton's Navigation and Piloting*) put it succinctly: "Location is time." And yet for many years honey bees were thought to ignore time altogether, measuring distance in terms of effort instead. Tests showed that foragers overestimate distance somewhat when flying uphill or into the wind, or when burdened with a 55-mg lead weight or a resistance-increasing flap glued to the top of the thorax. Flying downhill or downwind led to an underestimate of distance traveled. All of these are suggestive observations, though not conclusive.

Another alternative is that animals measure visual flow, the movement of objects streaming past as the creature walks or flies through the environment. "Objects" in this context actually refers mainly to the textured ground. Some versions of this idea allow the creature to ignore time. We will look at this possibility—the current favorite for bees—in more detail presently. Still another option, for birds and mammals at least, is to use information from the vestibular system, the elaborate organ complex in the inner ear that measures the direction of gravity as well as changes in speed and direction (acceleration). In this case, though, the timer errors are doubly serious: integrating acceleration over time gives speed; speed multiplied by time gives distance. Any error in judging time is thus factored in twice.

The most intriguing candidate for a time-independent distance computation strategy involves simple counting. For example, a recent test with blindfolded humans doing dead reckoning (on foot, traversing about 100 yards) showed that though the subjects could judge time only to 15%, they knew their distance to 8%. This is quite impossible if they are using the same internal clock and calculating distance as sailors do from speed and time. The vestibular system—using time twice—should yield even worse results. Perhaps, the investigators suggested, we simply count steps. Another group of researchers separated the possibilities by comparing blindfolded people walking (vestibular and counting systems potentially at work simultaneously) and pushed the same distance on a cart (vestibular system only). Active walking was more accurate, which implies a role for counting.

Humans on foot, it seems, may be measuring distance directly, as though we have a separate neural module for the task—calibrated, of course, to the length of our stride. If an animal has a default cruising speed—15 mph for honey bees, for instance—it is necessary only to count strides or wing beats or tail strokes subconsciously to derive distance independent of time. There is good reason to think that mice, fiddler crabs, and some species of ants do in fact count their steps. Fiddler crabs venture from their inconspicuous burrows along a circuitous scavenging route; when startled, they can do the trigonometry and run directly back to the unseen entrance of their homes. The distance-on-each-leg part of this computation could involve energy expended, visual flow, or even integrated acceleration, but in fact they count their strides. The key to the test was to put a piece of acetate under part of the route out. The crabs' scuttling steps on the slippery material covered very little ground; when the crabs set off home they overestimated the distance they had traveled going out.

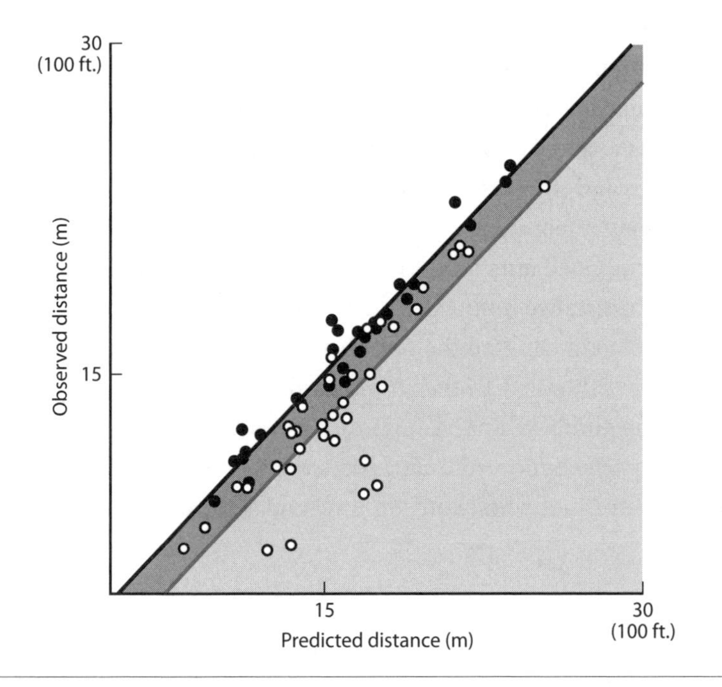

Distance measurement in crabs. During excursions from their hidden burrows fiddler crabs were displaced and their attempts to find their burrows were recorded. When the surface was solid and stable the crabs ran back in the apparent homeward direction for almost exactly the correct distance (dark circles; the solid diagonal line indicates results expected from perfect inertial navigation). When the surface was slippery (open circles) the crabs walked the appropriate number of steps but did not get as far. Note that errors (scatter about the best-fit line) do not increase proportionally with distance (time in transit) under either condition.

The most elegant test to date of the step-counting work-around involves desert ants. Because of their challenging habitat this species does not bother to lay odor trails or learn landmarks, as other ants do. There are few landmarks in the Sahara, and the blazing noonday sun destroys pheromone molecules quickly; even when the sun is low in the sky the next gust of wind will cover or carry away any odor-marked sand. Like crabs, the ants can navigate directly back to their colony entrance regardless of how circuitous the outward journey has been.

When a group of foraging ants was at a food source, experimenters glued pig bristles onto the legs of one set and partially amputated the legs of others. The bristle-legged ants took giant steps home and overshot the entrance; the leg-trimmed ants took tiny steps and stopped too soon. When these ants were returned to the colony their next journey was perfectly normal: once the bristle-equipped ants found food they were able to march the appropriate distance home; the short-legged ants were equally accurate. Each had counted the outward steps, done the trigonometry and, now calibrated to their new stride lengths, accurately computed the number of paces home. Chances are that many (perhaps most) species ignore the anthropocentric nautical solution and rely largely if not wholly on counting and self-calibration.

■ Going with the Flow

Keeping track of steps on the solid earth is one thing, but correcting for wind or currents seems a daunting challenge for aerial or aquatic animals. Once again, our intuition suggests a straightforward time- and distance based solution, but in fact we know that bees at least are doing something different. Bees automatically maintain a relatively constant 15-mph ground speed regardless of the wind, at least in gentle breezes up to half their cruising speed. They only slightly overestimate distances flown into the wind, so they must be able to compensate for much of the discrepancy. Flying a constant airspeed is about as much as we could reasonably expect of an insect; a particular amount of energy translates simplistically into a corresponding velocity through the air. Measuring airspeed is a straightforward matter of monitoring airflow. In the case of honey bees pressure on the antennae and the bristles embedded in the compound eyes seems to be key, since trimming either on one side prevents foragers from flying in a straight line.

Maintaining a set ground speed is far more complex. It requires measuring and allowing for the extra effort needed to overcome headwinds, detecting and compensating for side winds, and taking into account the extra push from a tailwind. Because wind speed cannot be sensed directly while flying (in theory at least, only airspeed is detectable), an animal would have to watch the ground and use the discrepancy between expected and observed motion to infer wind speed and direction, and from that work out the actual ground speed. For example, in a headwind at a fixed altitude the landmarks below will move by more slowly; in a crosswind movement will not be aligned with the body axis.

Compensation by human pilots requires geometrical calculations based on measuring angular movement over the ground and knowing the height above the ground. Like many flying insects honey bees are exquisitely sensitive to angular motion, particularly to the apparent movement of textures and features below them. One dramatic way to demonstrate this dependence on visual flow is to train bees across a body of water. If you string a row of floating landmarks along the route, the foragers fly normally. Remove the landmarks on a day when there is a lot of chop in the water and you will notice a bit of side-wind drift, but the foragers will manage the crossing. On a day when the water is calm and flat, enormous problems arise. The bees fly ever lower to resolve the texture they know must be there, completely failing to compensate for side winds and often descending so low that they strike the surface.

With a textured surface, bees have half of the solution: keep the angular motion at the right rate for your height above ground and your ground speed will be what you are aiming for. The problem for bees lies in actually knowing their altitude. To a low-resolution eye, flying at a ground speed of 12 mph at a height of 20 feet looks the same as a ground speed of 6 mph at an altitude of 10 feet. A bee, then, must have to know its elevation. Could bees estimate height by using binocular discrepancy—our own technique for judging the distance to nearby objects? Honey bees typically fly a zigzag path about 12 feet high. For an eye separation of 3 mm, this corresponds to an angular difference of about 0.04°. Because the resolution of the bee eye is about 1.5°, they clearly cannot be judging altitude this way. Birds at modest elevations, on the other hand, should have less difficulty using this approach.

The most intriguing suggestion for judging altitude (at least in insects) is that they might monitor the local horizon. The higher the animal is off the ground, the lower the horizon appears to be. We are not talking about the true horizon; for animals with compound eyes the change with height is far too small to detect (though birds should be able to use this strategy). Instead, fruit flies can use a nearby local horizon—the tops of bushes, for instance—as a guide. For objects this close, the parallax effect from changing altitude is substantial. Of course this only allows the creature to maintain a consistent relative altitude, not to judge absolute height.

Whether honey bees or other heavy-duty airborne commuters are equipped to judge altitude (and thus ground speed) in this way remains to be discovered. And while they obviously have clocks and timers to schedule their lives and orient to the sun, there is no compelling reason to think that they use time for determining distance, and good evidence that they avoid it for computing direction whenever possible. They have found one or more clever workarounds, only one of which (unconscious counting) seems to be in

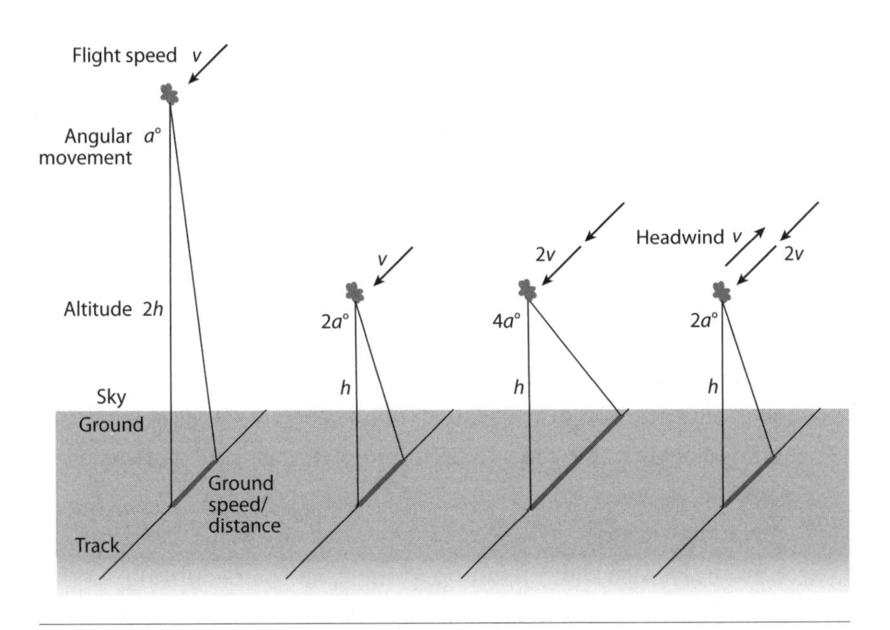

Speed measurement ambiguity in bees. A flying insect could measure its speed over the ground by solving a simple trigonometric problem: the distance traveled per unit time is the sine of the angular movement of the ground multiplied by the altitude. The bee in A senses an angular movement of *a*°, while the one in B measures 2*a*°, yet both are flying at the same airspeed and ground speed. The bee in part C is flying twice as fast and senses an angular movement of 4*a*°. The bee in D also is flying twice as fast but into a headwind; it measures 2*a*° over the surface below and yet has the same ground speed as in scenarios A and B. Without being able to measure altitude, judging true ground speed and compensating for the wind seems impossible. Nevertheless, bees manage to fly a relatively constant ground speed and maintain a fairly steady altitude.

our own repertoire. But like it or not, circumstances frequently conspire to force a bee to know the time of day—or more precisely, the interval since its last clock synchronization. So it is for navigating animals in general. It is to this question of calibration, and its reliance on celestial landmarks, that we now turn.

■ The Clockwork Universe

As we've seen, many animal navigators need to know one way or another, whether directly or indirectly, the time of day, the day of the year, the phase of the moon, the point in the tidal cycle, as well as the temporal duration of any current intervals being measured. To judge where they are in any of these cycles with maximal accuracy animals must synchronize their internal clocks to times, dates, and phases as often as possible. Human navigators face the same challenges, and our solutions to the problems suggest where we should begin to look for the techniques animals use in calibrating time.

Everyday experience suggests that we live on a static, fixed surface around which the sun, moon, planets, and stars dutifully circle. After all, if the earth were rotating, common sense tells us we should feel a constant wind blowing from the west. Given that the earth's diameter is about 8000 miles, early mathematicians calculated that this breeze would be a stiffish 700 mph at 45° latitude. Worse, if the planet did orbit our star, the speed around the sun would be about 65,000 mph—movement we could not easily overlook. Obviously we must be standing still; everything else must be traveling around us. And it must look this way to animals, though the larger philosophical questions of who orbits whom presumably do not arise.

In fact, however, we *are* in motion, and happily for life on earth the planet's atmosphere has sufficient momentum to move along with us. The key thing for navigators is that from our perspective the sun does indeed appear to circle the earth once a day, while the constellations seem to orbit us once a year. But the fact that we do revolve around the sun, and orbit along an ellipse rather than a circle, means that the times and motions navigators must deal with are not simple.

Like the sun and stars, the moon and the tides too have unnecessarily complex rhythms. Though they have precise average periods (27.32 days and 12.42 hours, respectively), the elliptical orbits and the interaction of the gravitational pulls of sun, earth, and moon as the distances between these three bodies continuously change generate substantial irregularities. The task for animals and humans alike is to predict the apparent motion of sun, moon, and stars, and use these celestial landmarks for inferring time, direction, and location. Along with the tides, these are the periodicities to which animal timers must synchronize themselves.

To talk about these temporal rhythms we need consistent units. We cannot know how animals divide things up, but some 4000 years ago the Sumerians chose to section the year into 12 months of various lengths, the day into 24 hours, the hour into 60 minutes, and the minute into 60 seconds. This arbitrary arrangement has somehow survived the world's love affair with decimalization. The Babylonians enshrined the Sumerians' sexagesimal bias when they chose to divide the globe into 360° of longitude and latitude; perhaps they selected this value because the year is about 360 days long. And then to the consternation of succeeding generations they sliced each degree into 60 minutes of arc, and each angular minute into 60 seconds. Minutes of arc and minutes of time only sound alike; the sun at the equinox in fact sweeps over 15 minutes of arc at the equator during each 1 minute of clock time. In fact, the original term was "primary minute"—that's "my-newt" with long vowels and a different emphasis, meaning very small, although the pronunciation today mimics the unit of time. Minutes of arc were then subdivided into "secondary minute" units, which we now (alas) call seconds of arc. Animals, of course, are not fooled. Their sun travels along its arc in the sky at a steady 15°/hour. The exact position of that path, however, changes each day.

Early astronomy was mostly about time. The positions of the heavenly bodies appeared to be the visible hands of a mysterious clockwork machine; their relevance to navigation was of little initial importance. That changed quickly as sailors discovered how to use the stars, planets, and sun as compasses—though to do so, they usually had to know the time of day with some precision, which was the astronomers' top priority in the first place. Portable and more practical versions of the telescopes and sextants that astronomers eventually used to calculate the position of the stars became standard equipment on ships, though they were forbiddingly difficult to use on a heaving deck. But measuring the position of a heavenly body is actually the easy part.

Calculating time and the movements of the celestial timekeepers depends on deducing their exact rhythms. This requires knowing present positions and being able to predict future coordinates. Calculating the expected movement would be easy if the sun and planets did indeed circle the earth. But of course even the moon does not rotate around the earth in a circle; instead it traces an ellipse. The sun and planets appear to trace wildly noncircular routes about us, accelerating and slowing down, approaching and then moving farther away. Only the stars behave in a truly aesthetic and orderly (i.e., circular) way.

Lacking computers, astronomers needed a set of relatively simple formulas to be able to use these celestial beacons as clocks and to anticipate their paths through the sky. For mathematical convenience they assumed that all objects except the seemingly motionless earth moved in perfect circles on transparent crystal spheres. This could never be made to work because in fact everything but the moon orbits the sun and worse yet, as Kepler first discovered, moves in an ellipse. The geometrical work-around invented 1900 years ago by Ptolemy had the sun moving in a small circle around a point on a larger circular path. Even this did not quite work, so later astronomers added more and more circles to the original one.

There is no chance that animals are programmed to predict the sun's position in this rococo fashion. And yet, "understanding" the odd motion of the sun *is* critical. And for many aquatic and littoral species the movement of the moon is very important too: not only is the position of the moon in the sky correlated with the all-important high and low tides, but the cycles of full and new moons predict the spring and neap tides, which can be key times for reproduction. As we will see, there is good evidence that the moon can be used as a compass.

■ The Sun and Stars as Clocks

While the planets trace impossibly complex paths in the sky, the sun and stars provide a less imposing mathematical task for celestial navigators. The stars are the surest and simplest guides; they are so far away that the earth's changing orbital position does not much matter, at least in the short run. The constellations appear to rotate through the heavens in perfect and complete 360° circles every 23 hours, 56 minutes, 4.1 seconds. This is the *sidereal day*—that is, the time it takes for a star to "circle" the earth once. If you do the math, multiplying the 3.94-minute shortfall from the solar day of 24 hours by the 365.24 days in the earth's annual cycle, you get the one extra day in the sidereal year. This discrepancy between the planetary and sidereal year arises because the earth orbits the sun. Most time systems arbitrarily start the day at midnight, when the observer's position on the earth is facing directly away from the sun. A star directly overhead at midnight on, say, the winter solstice will be directly overhead at noon six months later on the summer solstice. This means that the summer sky, at least at low elevations, has a different set of constellations compared to the winter sky. It also means that on average the stars "rise" about 4 minutes earlier each day. Both facts create complications for animals trying to orient at night, but at the same time provide at least in theory a way to calibrate annual clocks and, allowing for the daily drift, circadian timepieces as well.

A number of irregularities in the earth's motion can complicate use of the star compass. The earth spins like a gyroscope on its north–south axis, which is pointed (at the moment) within about 0.7° of Polaris. But the earth is not a very satisfactory top; it is not

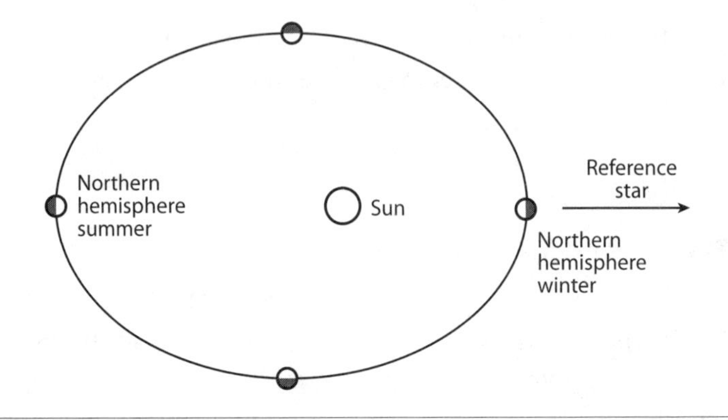

The sidereal year. The discrepancy between the sidereal and planetary year arises because constellations overhead at midnight in the winter (the "reference star" in this illustration) are then overhead at noon in the summer. Relative to the sun the earth rotates 365.24 times as it completes its orbit; relative to the stars, the earth rotates 366.24 times. As a result the star patterns appear to precess through the sky, so that low-elevation summer constellations are replaced by winter constellations as the seasons change.

perfectly symmetrical either in shape, distribution of land masses, or internal composition. Moreover, it has a heavy liquid core shifting its center of gravity in poorly understood ways, oceans sloshing about according to tidal rhythms (with dramatically more or less water depending on our current position in the glacial–interglacial cycle), and a moon whose own orbital peculiarities and consequent anomalies in gravitational pull affect the spinning of the earth. As a result there are 6-month, 14-month, and 14-year wobbles, a slight *nutation* or variation of the earth's axis of tilt every 18.6 years, a major precession of about 45° with a 25,800-year period, a minor cycle superimposed on the precession of 2.4° every 42,000 years, and longer cycles as well. The precession means that Polaris was about 45° away from true north only a few thousand years ago. Without recalibration even a 1° change in star position translates into a location error of about 70 miles; many species are much, much better than this. Hardwiring Polaris into the navigational equipment of animals is thus an unlikely strategy to expect natural selection to have employed. Animals need to learn the star patterns if they are going to use them and ignore the wandering planets. Once these have been memorized, changes in the earth's axis of tilt will generate errors of only about 0.3 miles over the lifetime of even a long-lived species.

Calibrating annual and daily rhythms to the sun is equally practical in theory, though the patterns of movement are significantly more complicated than those of the stars. The earth orbits the sun along an ellipse; we are about 7% farther from our star in the Northern Hemisphere summer than in winter. The earth's elliptical orbit means that the sun appears to revolve around us in a corresponding ellipse. When we are farthest away the earth is moving more slowly through space than when we are closer. As a result, the sun appears to drift out of phase with the rotation of the earth. But the largest complication is the earth's tilted axis, which (at the moment) is angled 23.44° relative to the plane of our orbit.

The axial tilt is the source of the seasons, which provide the impetus for most animal migration. Many migrators move north in the spring to take advantage of the longer days for foraging (as well as the paucity of predators and parasites, killed off by hard winters). The seasons and latitudinal zones they are exploiting are defined by the earth's tilt. The Northern Hemisphere is angled toward the sun in the summer and away in winter. On June 21, the summer solstice, the sun passes directly overhead at noon along the Tropic of Cancer, at 23.44° north latitude. On September 21 and March 21 (the equinoxes) the sun is overhead at noon on the equator, and on December 21, the winter solstice, the sun is overhead at noon along the Tropic of Capricorn (23.44° south latitude). Thus, the tropics are defined as the region where the sun passes directly overhead once or twice during the year. By contrast the polar zones are the latitudes above the Arctic and Antarctic circles, where the sun does not rise at all for at least one day of the year and does not set for at least one day: 66.56–90° north and south latitude. The temperate

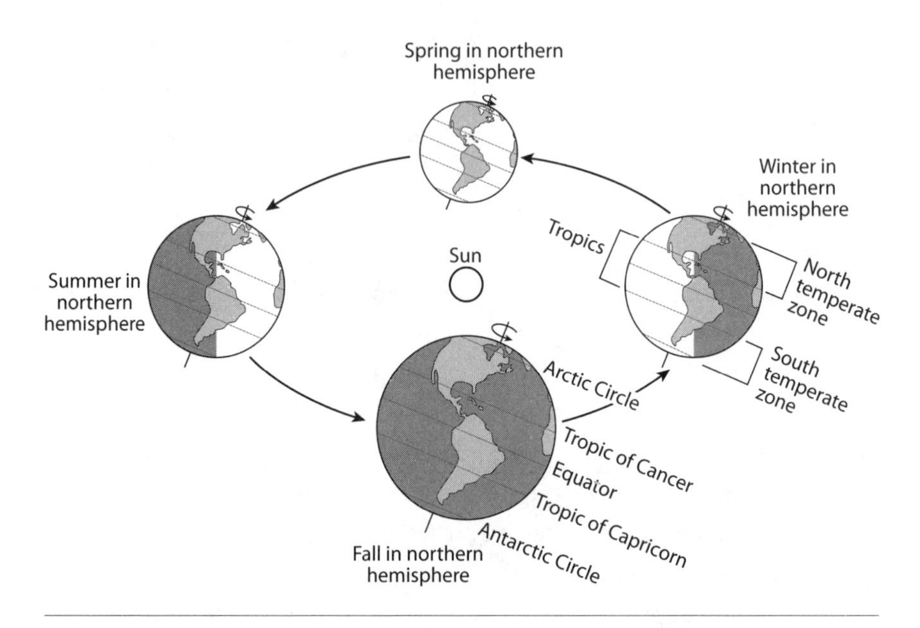

The sun and the seasons. Because the earth is tilted on its axis relative to its orbit around the sun, more of the Northern Hemisphere is exposed to the sun during part of the year (left) than six months later (right). These alternative periods of long and short days correspond to summer and winter. This tilt of 23.44° defines the Arctic Circle (the latitude at which the sun does not rise on the winter solstice: 66.56°) and the tropic lines (latitude 23.44°) where the sun passes directly overhead on the summer solstice. The earth's orbit is an ellipse; the earth is farthest from the sun during the Northern Hemisphere summer.

zones are the regions in between. Earth's champion navigators migrate from summer breeding grounds near the Arctic Circle to wintering quarters in the warm tropics (or beyond). Bar-tailed godwits, for example, enjoy an endless summer; they breed in Alaska during the Northern Hemisphere summer, then fly south, crossing the equator, and spend the other half of the year (the Southern Hemisphere summer) in New Zealand.

The tilt of the earth has two immediate consequences. First, the intensity of sunlight is greatest directly under the sun's arc in the tropics. At noon the light passes perpendicularly through the atmosphere; this short path length minimizes absorption and scattering in the air, maximizing the energy reaching the surface. At other latitudes the path is longer, and thus more energy is dissipated in the atmosphere. The second factor is day length; under the sun's arc the day is exactly 12 hours long. Closer to the pole tilted toward the sun the summer days are longer, while at latitudes in the other direction the winter days are shorter. Though the total amount of light energy actually penetrating the atmosphere per hour is lower in polar regions, the days have more hours—up to twice as many as in tropical regions.

Not only can polar plants receive more net light per day, but that light can be more effective than the unfiltered light of lower elevations. Solar radiation in the tropics and even the temperate zone can overload the chloroplasts in leaves, leading to an actual loss of photosynthetic energy (*photorespiration*). This is less of a problem at high latitudes. Though polar regions never get as warm as tropical areas, the long days and wet thawing soil support a highly active community of plants and astonishing numbers of insects. The insects in turn nourish vast populations of nesting birds—songbirds, waterfowl, and shorebirds. Many commute over surprising distances to gorge in the Arctic: while most songbirds travel only a thousand miles or so to higher latitudes, or up to 3000 if they vacation in the tropics, the amazing bristle-thighed curlew migrates nonstop to Alaska from its wintering grounds in the South Pacific, a trip of more than 6000 miles. And curlews are relative homebodies compared to the arctic tern, which covers about 27,000 miles annually, breeding in the northern tundra and spending the Northern Hemisphere winter feeding at sea off Antarctica during the Southern Hemisphere summer. The climate-altering tilt of the earth motivates most of the long-distance migratory journeys on the planet, journeys that require anticipating the changing angle of the sun in order to set off north and then south at just the right time.

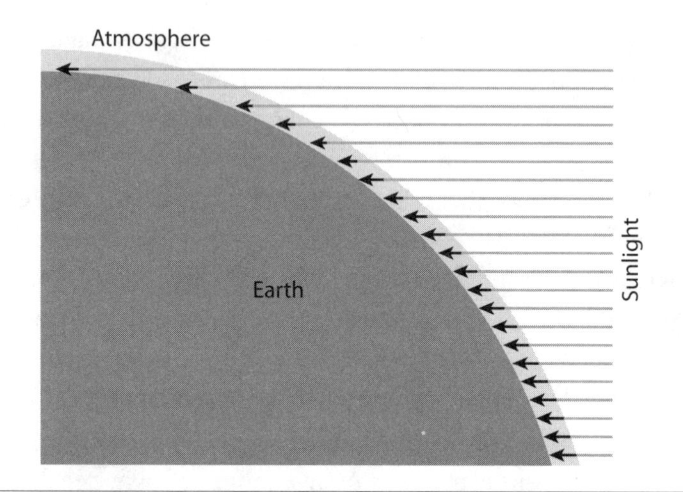

Path length through the atmosphere. Sunlight is absorbed and scattered in the air. The degree of energy dissipation is lowest directly under the sun's arc at solar noon. This diagram can be taken as a noontime side view of the earth, with greater path lengths at more northerly latitudes. Alternatively it can represent a top view showing the variation in intensity under the sun's arc over the course of a day; near dusk (top) and dawn the sun's intensity is greatly attenuated because of the greater path length the light must traverse.

■ Synchronizing the Year

Wherever they go, the problem for migrating birds is to time their journeys to arrive as early as possible and stake out the best territories. Arriving too early, however, may mean starving or freezing to death; birds also perish en route when weather and food supply conspire against them. Departure date in the autumn is less critical, though still important. Migrants must leave early enough not to be caught in bad weather along the way, but only after lingering in the insect bounty long enough to recover from rearing their brood and to pack on enough fat reserves to make the journey. Spring or fall, their main cue is the changing length of the day or night.

The date in the annual cycle at any temperate latitude is rigidly correlated with the number of daylight hours—assuming you know your latitude and are equipped with the appropriate formulas for computing the date. Experiments in the lab suggest instead that animals calibrate their calendars once or twice a year in some latitude-independent way. The most obvious but least useful cues are the extremes; the longest day is the summer solstice, the shortest the winter solstice. Though a solstice might seem important in synchronizing the annual cycle, solstices generally occur months before or after animals need to set off.

More useful for calibrating a creature's endogenous annual clock are the equinoxes. Not only is the day exactly 12 hours long

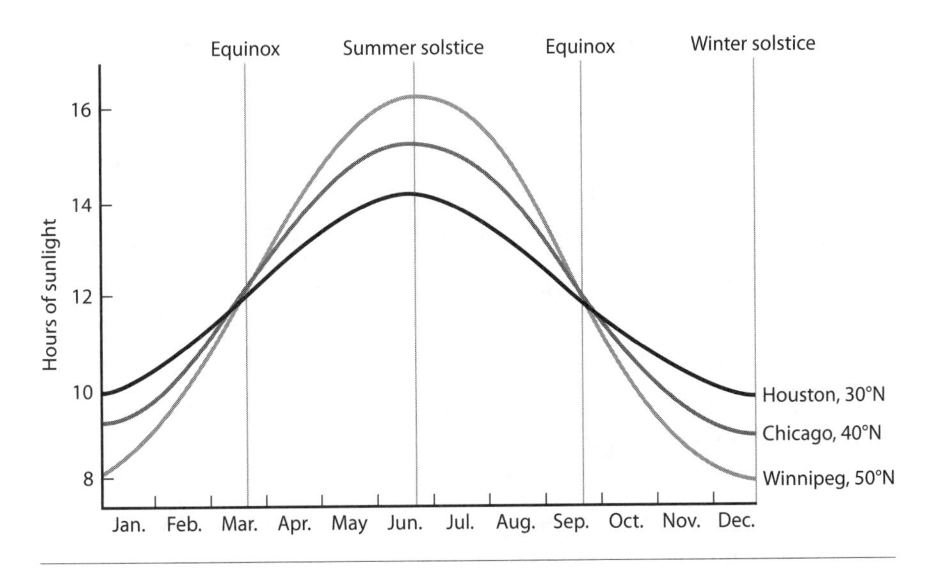

Latitude and the day–night cycle. As an animal in the temperate zone moves away from the equator, the length of the days in summer and nights in winter become ever more extreme. (As is evident from the point at which the curves cross, the day is 12 hours long on the equinoxes regardless of latitude.) At the same time, the exact length of the day changes very slowly at the solstices but relatively rapidly near the equinoxes, as indicated by the slopes of the curves.

on March and September 21 *regardless* of latitude, the daily change is at its fastest—several minutes a day. Noticing the difference between 11 hours 56 minutes and 12 hours 00 minutes has to be easier than judging whether 13 hours 18 minutes is longer or shorter than 13 hours 19 minutes. Early humans built special structures like Stonehenge to determine these four special dates—the solstices and equinoxes—and then counted off days until the next resetting. These temples of time had a huge advantage: they were fixed to the ground, and permanently oriented relative to true north. An animal, particularly one on the move, has to carry a portable Stonehenge.

Outside the temperate zones in the tropics and polar regions, calibration is still more difficult. Because many migrants spend most of their lives in one or the other, merely commuting through the temperate climes, they must deal with the resulting complications. On the equator, for instance, day length varies by only two minutes over the entire year; from one day to the next the difference is 1.5 seconds at most. In the polar regions there is an analogous problem with day length in the summer: the sun does not set at all for up to six months. Another problem at high latitudes is that using the 12-hour day of the equinox to reset the annual clock is hard because sunrise and sunset are difficult to judge. The reason is that the sun's arc through the sky is very shallow, reaching at the pole a maximum elevation of just over 23°, just skimming the horizon for hours at a time.

So how can migrating animals hope to calibrate their annual clocks in the tropics and polar regions? They might use the stars. Or perhaps they wait for their transit through the temperate zone, where 45° latitude will provide the best signal. Another possibility is that in the tropics animals use the annual pattern of wet and dry seasons, a cycle driven by the earth's tilt and the corresponding rain bands that the periods of perpendicular solar radiation create. If animals know their latitude (and this information is in the elevation of Polaris and in the earth's magnetic field, suitably corrected) and understand the pattern of noontime elevations of the sun, they could conceivably compute the date. If errors in these estimates are symmetrically distributed, as in theory they should be, repeated daily measurements could increase accuracy as random errors slowly cancel one another. In fact there is no one solution to this problem; different species have hit upon different tricks, only some of which are beginning to be deciphered. But all systems use multiple measurements to enhance accuracy. (Alas, this is a case of diminishing returns: precision increases only with the square root of the sample size, so that doubling the number of measurements reduces random errors by only about 30%.)

■ Synchronizing the Day

While synchronizing the annual cycle at least once a year is crucial for migration and breeding, recalibrating a circadian clock needs to be done daily. The clock then ticks off the minutes, allowing a bird, for instance, to know when to start its singing in the dawn chorus, what time to start foraging (when its prey become active), when to begin worrying about overheating at midday, as well as to anticipate dusk and thus begin to wind down from its hectic schedule. This daily calibration seems, at first glance, a simple task. Outside the polar regions the sun gives us a fairly sharp dawn and dusk, and reaches its maximum elevation due south at solar noon, exactly halfway between rising and setting. But the time of sunrise in the temperate zone can change significantly from one day to the next (significantly, that is, if you are using time to help navigate), and judging solar noon is difficult because the sun's elevation changes quite slowly at midday. Worse yet, all of these parameters depend on latitude. Take, for example, a day on which sunrise at 40° north (roughly the latitude of New York City and Rome) is at about 4:45 a.m. and at an azimuth of 60°. (All azimuths are measured clockwise from 0°, true north; thus east is 90°, south 180°, and west 270°. Times are solar, so that noon is the moment when the sun is due south.) The sun would set at 7:15 p.m. at an azimuth of 300°. On the same day at 60° north (approximately the latitude of Anchorage and Oslo) the sun would rise instead at 2:20 a.m. and set at 9:40 p.m. at azimuths 35° and 325°, respectively. Clearly latitude matters enormously. But so does date. On the winter solstice the sunrise and sunset at 40° latitude are at 7:45 a.m. and 4:15 p.m. At latitude 60° these times are 9:15 a.m. and 2:45 p.m. Humans need a thick volume of tables to sort this out.

Seamen in the age of sail made a ritual of setting the ship's clock to solar noon by determining the moment of highest solar eleva-

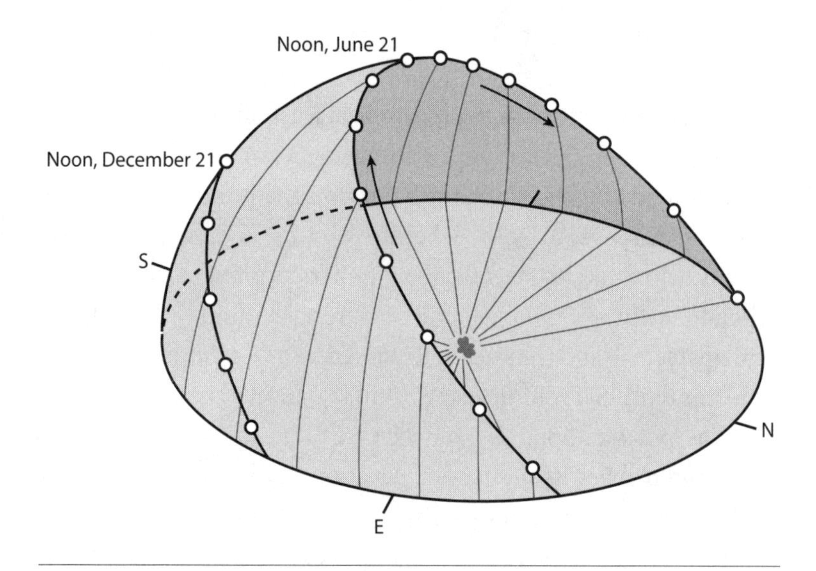

The sun's arc at 40° north latitude. The path through the sky is shown for the summer and winter solstices. On all other dates the arc is parallel to these two trajectories but somewhere in between. The time of dawn and the maximum elevation of the sun are different each day. The consistent features are that the sun is in the south and at its maximum daily elevation at solar noon, noon occurs exactly halfway between dawn and dusk, and the angle between the rising sun and the noon sun is identical to the angle between the noon sun and the setting sun.

tion. This was an inexact process even when the sky was clear, the sea calm, and the winds moderate. Experience proved that another strategy gave more consistent results: they would time the interval from dawn to dusk on one day and then use half that time to judge noon on the next day. In temperate latitudes the error is rarely more than a minute, and an affordable chronometer more accurate than that was a rarity until well after 1800.

A major problem for sailors, and presumably animal navigators as well, is that the horizon is not always visible. Beyond the obvious problems of clouds that hide the rising sun and hills that change the elevation of the horizon (rarely a difficulty at sea), the ocean air is often hazy at low elevations even when the sky is clear. Another worry, particularly for birds, is that the elevation of the observer matters. A lookout at the top of the mainmast of a ship sees the sun rise a few moments before the crew on deck; the resulting difference is sufficiently great that every good navigation text provides formulas for correcting this parallax effect. For an animal at 1000 feet (and migrating birds are regularly found at ten times this elevation) the discrepancy is stupendous, and would thwart any very precise attempt to reset the daily clock without some built-in strategy for compensation.

By the 1700s sailors and other navigators had begun to rely on elaborate tables and exact measurements of distances between the ever-moving moon and various stars in the background. This could be used to judge time, and thence (laboriously) to calculate longitude before accurate oceangoing clocks were available. These lunar measurements unfortunately depended on clear weather and the presence of the moon above the horizon.

We have not considered the recalibration of lunar and tidal clocks. Most current evidence suggests that the sight of the full moon—probably its position overhead in the sky at midnight—and the maximum water pressure on submerged marine creatures at high tide are the most important stimuli for synchronizing these clocks. These two cycles (with a specific delay after a full or new moon) allow numerous species to anticipate spring tides, when the moon and sun are aligned so that their combined gravity exaggerates tidal fluctuations; they also predict neap tides, when the difference between high and low tide is at its minimum. In addition to timing migration and mating, these tidal cycles are crucial in predicting when shorebirds can find the richest concentrations of prey stranded by retreating waves—and when prey animals should swim back or dig like mad into the mud to escape being eaten.

We've focused in this chapter on timers as a way of knowing the date in the year or lunar month, the hour in the day or tidal cycle, and the interval between two events or measurements. Beyond specifying when to begin migrating or mating, the primary use of timers is to allow animals to determine direction, correcting for the movement of compass cues like the sun and the stars. It is to this time-compensated compass challenge that we now turn.

Chapter 4

Insect Compasses

Anyone who has dug in the wet sand near the tide line will have encountered sandhoppers, tiny crustaceans that hop wildly when threatened. Sandhoppers are usually out foraging only at night, feeding on detritus the waves have left behind, having anticipated both dusk and the ebbing tide. In preparation for an incoming tide or dawn they burrow back down into the sand.

The approach of a predator makes them flee toward the safety of the ocean. In the daytime, unearthed from their hiding places by foraging shorebirds turning over the sand for worms and shellfish, they make a run for the water. Sandhoppers also migrate periodically to the water to mate, a behavior that requires them to predict the monthly spring tide and understand the time of year. Clearly, they must know pretty precisely the orientation of the sea–land axis; but since much of their behavior occurs in complete darkness, how do they keep track of it?

In the lab, tests on these unimpressive creatures reveal remarkable navigational programming. First, they know the direction of the water innately. Sandhoppers from the north–south coast on the Atlantic shore of Spain flee west; those born on an east–west beach on Spain's Mediterranean coast hop south. These preferences are evident even if the mother was transported elsewhere while carrying her eggs: the young of a mother from the Atlantic coast, though born on the Mediterranean, nevertheless flee to the west. What cues do they use, and how can the newly emerged sand-hoppers know which direction is west or south without first calibrating their sources of information? Experiments show that they use the slope of the shore to establish the local axis bearing, define this as the innate direction specified by their genes, and then match to it the sun's movement, the rotating polarization patterns, the moon's travels, and the earth's magnetic field, shifting effortlessly from one cue to another as the situation warrants.

Like most navigating animals, these small invertebrates must understand something about current time, including the date in the annual, lunar, and tidal cycle, as well as time intervals. Time is critical for the cyclical activities of feeding and mating, but also for interpreting compass information, allowing in particular for the sun's daily journey from east to west. The problems are similar for vertebrates, though on a much larger scale. As a matter of intellectual convenience we will look at insects first, where journeys are more often local and experiments more easily controlled. We will examine vertebrate compass behavior in the next chapter.

■ Our Star

Humans are overwhelmingly visual animals. When we talk about our sensory or conceptual limitations the usual metaphor is blindness. When we grasp a concept or an argument, we are likely to say that we see the point. Darkness is synonymous with fear in a way that silence or the odorless world of anosmia can never be. Ignorant of what we are unable to perceive, we remain blind to our own blindnesses. Color blindness, which afflicts more than 5% of the population, went undiscovered until 1798, and only by studying insects did we awaken to the possibility that some organisms can see ultraviolet, infrared, and polarized light. We are blithely unaware of our lack of a magnetic or electrical sense. For centuries even the most far-seeing natural scientists took it for granted that animal navigation must rely first and foremost on visual cues—and moreover, on the range of visual cues that humans perceive.

Indeed the majority of species have at least some provision for visual orientation, most often using the sun as a compass. The sun is central to our world, generating light and life. Any number of ancient cultures have incorporated sun gods or sun worship, focusing on understanding the annual cycles of growth and game movement that either correlate with or are controlled by the sun. The arrival of spring, the summer solstice, the autumnal equinox, the end of the cycle of shorter days in midwinter, and so on were critical to human survival, and their prediction conferred civil or religious power on the predictor. As we saw in the previous chapter, animals have in some sense discovered the same regularities in the sky, and attune their annual cycles of migration and reproduction accordingly. Once humans noticed that animals can orient and navigate, it was natural that the first cue to suggest itself was our local star.

In the first chapter we divided orientation and navigation strategies into six categories; each of these can potentially make use of the sun as a guide:

- taxis, the most obvious and widespread form being phototaxis;
- compass orientation, using the sun as the primary reference cue;
- vector navigation, with the sun as the compass on each leg;
- piloting, with the sun as the necessary third landmark (not usual for sailors, but common among animals);
- inertial navigation (dead reckoning), using the sun to derive a bearing on each leg; and
- true navigation, with the sun as the primary compass to use in conjunction with a global map.

The sun may be the most plausible compass to the human mind, but it is a problematic landmark. Its azimuth changes from east to west each day, so that to use it in computing bearings over time requires compensating for this apparent movement in the sky. The exact rate of change depends on both the latitude and the date, as we have seen. The difference in azimuth movement near dawn and dusk on the one hand and noon on the other can be enormous. At 40° north on the solstice the rate varies by a factor of five. When the midday sun is sweeping from east to west at nearly 50°/hour the potential for navigational error is huge.

Many species live in the tropics where the situation is more confusing still, at least to us. In some parts of the year the sun's azi-

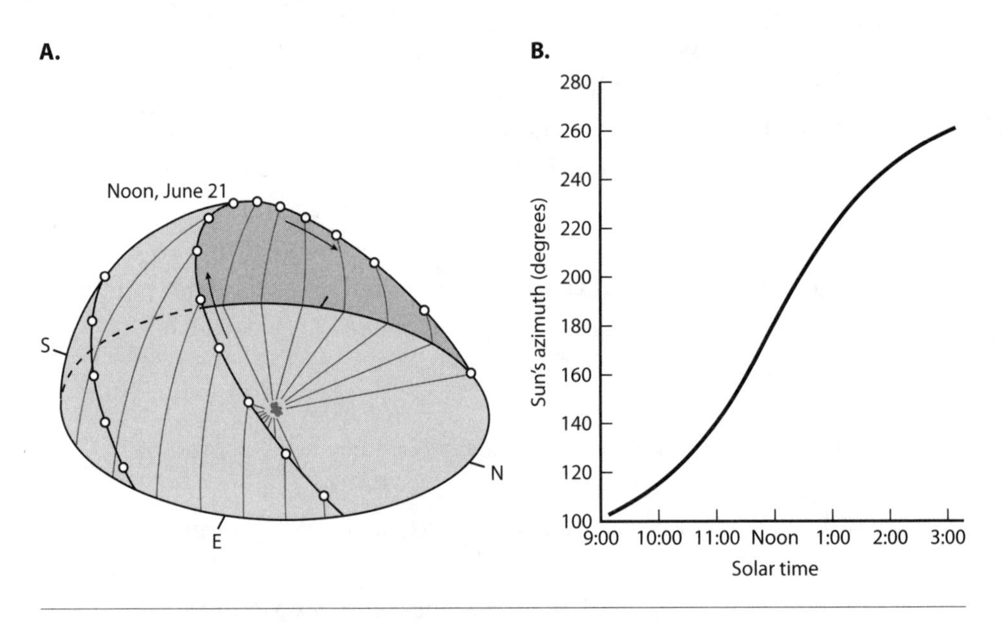

The sun's path through the sky at 40° north latitude. (A) As we saw in the previous chapter, on the summer solstice the sun rises early and climbs to an elevation of 66.4°. (B) The rate of azimuth movement varies over the course of the day, especially near the summer solstice at temperate latitudes. Near noon at 40° north the sun's azimuth is moving west at 47°/hour; at 11 a.m. the rate is 24°/hour; at 6 a.m. the figure is about 10°/hour.

muth moves clockwise (CW) from left to right (as in the northern temperate zone), whereas in other parts of the year the direction is counterclockwise (CCW), from right to left. The movement is always from east to west; the difference depends on whether at its maximum elevation at solar noon the sun is in the north or the south. Azimuth changes in the tropics can be very rapid in late morning and early afternoon. The most extreme case occurs at the equator on the vernal and autumnal equinoxes, when the sun rises in the east, climbs straight up for six hours at azimuth 90°, passes straight overhead (the zenith point in the sky) at noon, and then spends the next six hours at azimuth 270° (west) until it sets.

However daunting the task of allowing for the sun's continuous change of azimuth, the ability of animals to compensate for this wandering landmark has been clear since at least 1915. Four years earlier, ants en route from their nest to a food source had been redirected by a researcher who blocked their view of the sun and then provided a mirror image from a different direction. The ants turned in their tracks. When he moved the mirror to a different angle, their orientation changed correspondingly. True, the reorientation was not perfect, and some species seemed to ignore the sun, relying perhaps on odors or landmarks or other celestial cues instead. Still, for the ants that *were* fooled by the mirrors, the effect was dramatic. And then in 1915 foragers from one of the sun-compass species showed that they could allow for the movement of the sun after a period of confinement en route. Researchers captured them on their way to a familiar goal, kept them in the dark for a time, and then set them loose in an unfamiliar spot. Most of the insects adopted an angle relative to the sun that compensated for its intervening movement. How they did it, and with what precision, remained unexplored.

Another technique for demonstrating that a species uses the sun as a compass is the clock-shift experiment described in the previous chapter, when we were trying to infer the accuracy of animal clocks. Instead of asking when forager bees trained to gather food at a specific time do in fact turn up at a feeder, the researcher looks at an ongoing all-day behavior like migration. If, for instance, monarchs en route SSW to their winter haven in the mountains of Mexico are captured and clock shifted, their orientation when released will tell us whether the sun is their guide. In fact butterflies shifted by 6 hours fly 90° to the right of their shiftless peers, adopting the appropriate angle relative to the sun for the apparent time of day. If they used odors, magnetic fields, landmarks, or any of a variety of other potential sources of information as their primary guide, they would not have been fooled.

As the monarch example reminds us, insects, given their size and frequent willingness to ignore humans and equipment, can be very convenient subjects. Indeed, it took another 35 years to figure out how to ask the same question of birds in the laboratory and obtain consistent results. This pattern has repeated itself with several other orientation cues: an ability is discovered in insects, and then later turns out to be part of the (often more elaborate) navigational repertoire of vertebrates. In our discussion of animal compasses we will generally be repeating this historical sequence, starting with their discovery in insects.

■ The World as a Mosaic

In some ways insects have a simplified view of the world. Their compound eyes are composed of hundreds (ants) or thousands (bees) of individual facets (*ommatidia*), each a kind of zero-magnification telescope looking in a unique direction. Humans, by comparison, have a single lens that projects an image onto millions of photoreceptors. The exact resolution of compound and vertebrate "camera" eyes is strongly correlated with an animal's size. A fruit fly has a visual resolution of about 7°; a small bird 100 times larger (and 100,000 times heavier) can see about 350 times better—0.02°. For small body size, the compound-eye strategy can actually deliver a better resolution than would a camera eye of the same diameter, and it weighs quite a bit less. Like vertebrates, diurnal species generally have color vision. Many insects can see ultraviolet light but not red; some vertebrates such as homing pigeons are now known to see in the ultraviolet range as well. Most insects can see polarized light; inspired by this discovery, researchers have found a similar sensitivity in a number of birds, fish, amphibians, and reptiles.

While compound eyes have relatively low spatial resolution, they are extremely efficient at capturing light—100 times better than the eyes of most vertebrates. This efficiency can be used in different ways, most obviously to see in dim light. There are species of nocturnal bees, for instance, that navigate using landmarks illumined only by starlight; many owls have similar levels of sensitivity. Some bees can even distinguish colors under these unpromising conditions; owls apparently cannot. More often, though, all those extra photons are used to increase the "shutter speed" of the insect eye. When we watch a movie or television we see a continuous image; in fact, however, the movie screen goes dark 24 times per second while the next frame is moved into position between the lamp and the lens. On a television a broad black line moves vertically across the screen 30 times per second (25 times per second in countries with 50-Hz current). We are generally unaware of these interruptions because our eyes are able to distinguish only about 16 events per second. Our peripheral vision has a lower resolution but a faster shutter speed; some people can see fluorescent lights flicker out of the corner of the eye. For a diurnal insect the flicker-fusion rate is closer to 250 events per second—about the wingbeat rate of flies and bees. In short, most invertebrates have very high temporal resolution (and correspondingly quick reactions), an ability unmatched by any vertebrates.

While the high fusion rate is useful to flying insects in that it eliminates the blur in passing landmarks, the lowered resolution places an obvious limit on navigational accuracy. For a honey bee each facet covers about 1.5° of the world. Dragonflies have the best resolution (better than 1°), while some tiny insects with many

How a compound eye sees the world. (A) Honey bees have about 3000 facets in their eyes, and a total field of view greater than 180° horizontally. Each ommatidium covers about 1.5°. At close range this low-resolution picture nevertheless picks out all the essential detail. (B) The same hibiscus blossom at human-eye resolution. (C) A 180° panoramic view of a field and forest edge as a bee might see them. The sun is somewhere inside the bright white "pixel."

fewer ommatidia carve the visual world up into 10° or even 15° segments. So how good (or bad) is 1.5°? For calibration, the sun and moon are each 0.4° in diameter; 1.5° is roughly the width of your thumb held at arm's length. Obviously a bird should, in theory, be capable of measuring the sun's azimuth far more precisely than a bee. The camera eye also has a big advantage at night. While most species of birds are unable to distinguish shapes under starlight, they can make out the stars themselves and (as we will see) use them for navigation. For insects this is an optical impossibility.

While ants are convenient because they are easy to keep track of—they walk rather than fly—far more work has been done on honey bees. In part this is because the great Austrian biologist Karl von Frisch happened to choose them as his study animal and proceeded to unearth one remarkable ability after another. The economic value of the species has doubtless been a factor too. And it doesn't hurt that they are furry and cute, quick to learn, easy to keep, hardworking, and reliable—the border collie of the insect world. But the key advantage of the use of bees for navigation studies is that when foragers return to the hive they draw a miniature map to show potential recruits the location of food they have found. This remarkable communication system, the *dance language*, is the second most information-rich exchange in the animal world. (Human language is number one, by a large margin.)

■ The Dance Map

When foraging bees return from an especially rich source of food they often perform a waggle dance. This maneuver is carried out in the darkness of the hive on one of the vertical sheets of comb, usually near the entrance. During the dance the bee moves in a compressed figure-eight pattern, swinging its body ("waggling") during the inner parts of the cycle. The direction of the waggling runs encodes the direction of the food: using up as the direction of the sun, the orientation of the waggling left or right of vertical specifies the relative azimuth of the resource. Thus, if the dance is pointed 80° left (CCW) of vertical, the food lies 80° to the left of the sun's azimuth. The duration of the waggle run indicates the distance to the food. For the subspecies von Frisch studied when he discovered the dance language, each waggle corresponds to about 50 yards; for the yellow Italian subspecies so common around the world, the conversion is about 20 yards per waggle. For other subspecies the innate "dialect" can range down to five yards per waggle.

That the dance direction is keyed to the sun underlines the importance of celestial cues to honey bees. At first glance it's hard to imagine how this system could have evolved with its up-is-the-sun's-azimuth convention. In fact, there were two small steps involved rather than one large one. Our temperate-zone honey bee originally evolved in the tropics, where several other species of honey bees still live. The species thought to be most primitive—that is, most like the ancestral group—is the dwarf honey bee. It builds its comb in the open, with a broad, gently curved top. It is on this horizontal surface that the foragers dance; their waggle runs point directly to the food. Our temperate-zone honey bee will dance just outside the hive entrance on hot days if a sufficient crowd of potential recruits is assembled there. If the configuration of the entrance permits them to dance on a horizontal plane, the bees orient to celestial cues rather than to gravity.

The second step was to incorporate an odd but consistent insect behavior. Take an ant, for instance, running along a line 45° to the right of a light on a horizontal surface. Turn off the light and tilt the surface to the vertical and the ant will, without even breaking stride, randomly set off either 45° left or right of vertical. Now add selection pressure to suppress the left option, and you have the honey bee dance convention.

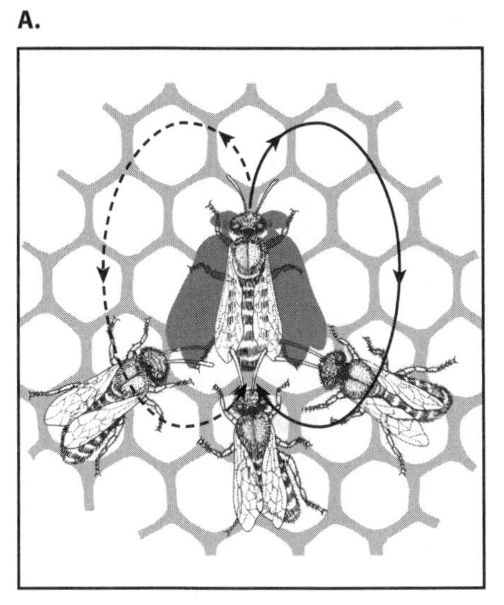

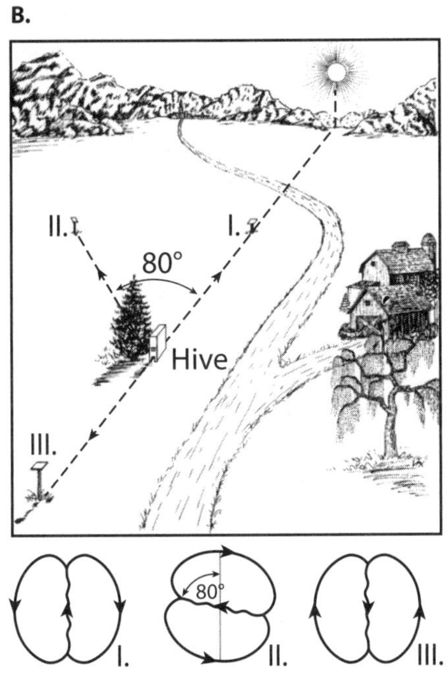

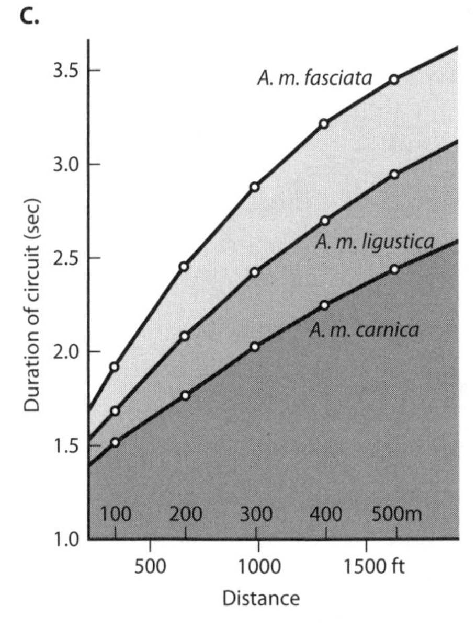

The dance language of honey bees. (A) The dance form is a compressed figure eight, with waggling during the straight central portion of the dance. (B) The orientation of the dance relative to vertical specifies the direction of the food relative to the sun's azimuth. In the three examples here, I is vertical, indicating the food is toward the sun; II is 80° left of vertical, corresponding to a resource 80° CCW from the sun's azimuth; III is aimed downward, which means the food is directly away from the sun. (C) The time spent waggling correlates with the distance to the food, but the exact conversion depends on the subspecies. The waggle rate is 13 per second.

The horizontal-dance holdover from perhaps a few million years ago provides a convenient way to ask our temperate-zone bees how they orient. If we put the colony in an observation hive (a single sheet of comb between two layers of glass) we can observe all of the dancing. The colony is necessarily small, perhaps a tenth the usual size, but in other ways behaves quite normally. If the hive is laid carefully on its side in the dark (or in red light, which is the same thing to the bees) some dancing continues, but the waggle runs are disoriented. If, however, the dancers are permitted to see the sun in the sky, the waggling points directly toward the food. If we block any direct view of the sun and substitute a mirror image, the dances rotate into alignment with the reflection. Do the same thing in a darkened room and show them an ordinary light and the foragers will orient their dances to the electric bulb as though it were the sun.

Honey bees, like all insects, have three problems to deal with before they can hope to use the sun as a compass. The first, oddly enough, is to identify the sun. We might think this is dead easy: it's the intensely bright disk in the sky. But there are potential problems. First, with a visual resolution of 1.5° bees cannot hope to actually see an object 0.4° degrees across as a distinct shape. As to brightness, the sun often appears as a distinct but dim disk through high overcast. On partly cloudy days the sun can be hidden behind a cloud while it illumines other clouds in the sky; often those illuminated clouds are the brightest spots in the heavens.

When we present the bees in a horizontal hive with various artificial suns, they reject some as obvious frauds and dance randomly. But they accept others even if they are the wrong shape or color, such as a green triangle 10° across. Dancing foragers make the sun/not-sun call based on an innate rule: if the bright spot shown to them contains no more than 20% ultraviolet light and is no larger than 15° across, then it is the sun. Otherwise, it's something else. This is a rule of thumb that works pretty well; ultraviolet

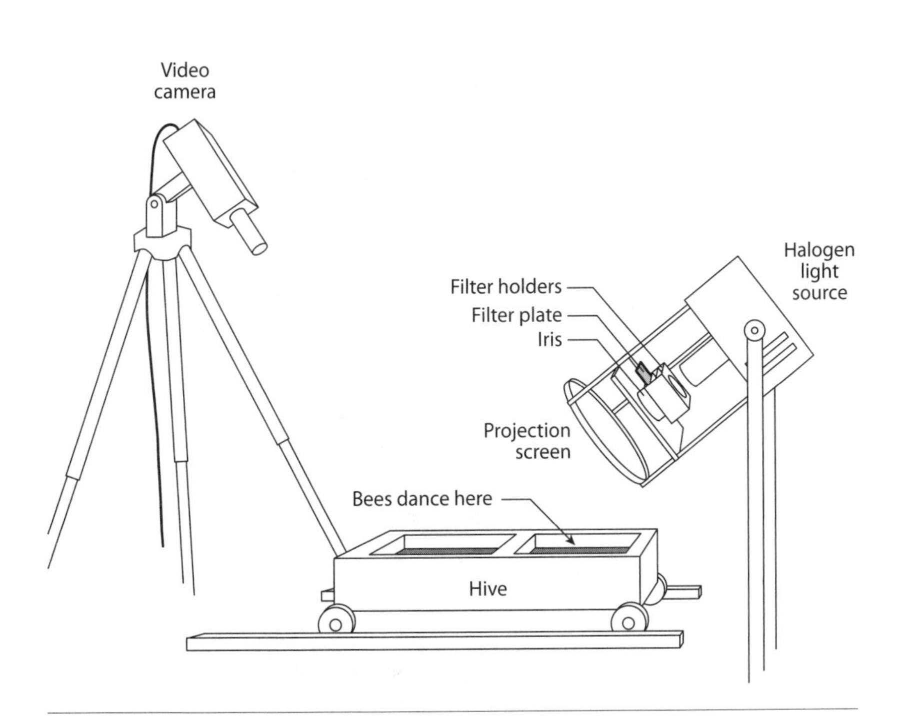

Horizontal hive. The observation hive is lying on its side in a darkened room. The artificial sun seen by the dancers is a quartz light source modified by various filters and projected on a screen (right). The resulting dancing is observed with a video camera (upper left).

light is mostly found in the blue sky, where it is the wavelength (along with blue) that is preferentially scattered down to our eyes by the air. Scattering is why the sky is not black like the outer space behind it, at least from earth; on the moon where there is no air to scatter the sunlight, the sky is indeed black. The blue- and ultraviolet-scattering bias of air is why the sky is not white like the sun's direct light. Of course if bees had large camera eyes, none of this would be necessary. (Small vertebrates on the order of an inch in length have marginal visual resolutions on the order of 0.5° or worse, but nothing is known about how they localize and identify the sun.)

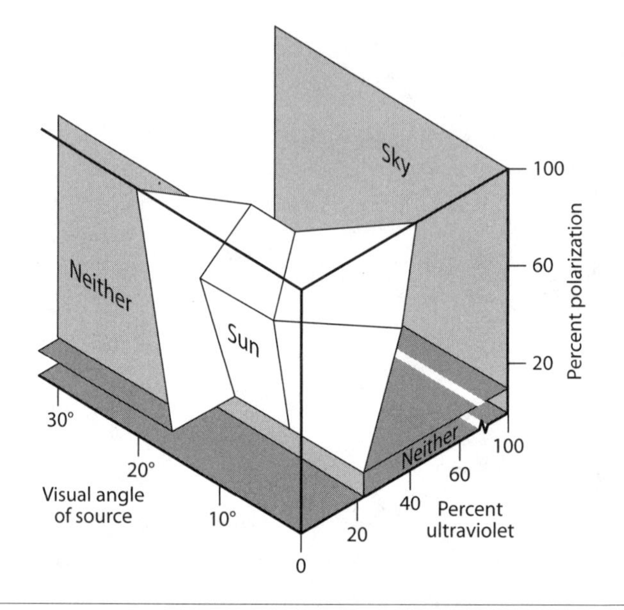

Recognizing the sun. Bees interpret some light patterns as representing the sun (the corner of the diagram that appears nearest) and other patterns as something else (the sky for instance). The decision is based on two simple rules: the pattern must be no larger than 15° across, and must have less than about 20% ultraviolet light. Degree of polarization does not matter, though in the actual sky direct sunlight is never polarized while the scattered light in blue sky is, especially well away from the sun.

■ The Moving Target

The second problem, faced by all animals whether they sport a backbone or not, is that the sun moves. We are not tackling the difficult question of its angular rate of movement at this point, but a more basic difficulty: Which way does the sun sweep across the sky? In the northern temperate zone the answer is left to right (CW); in the southern temperate zone, it is right to left (CCW). In the tropics the answer depends on latitude and date, and changes twice a year.

All species of honey bees can be found in the tropics, so we might expect that they have evolved to solve this problem. Many species of birds either live in the topics or migrate there for the winter, so this abrupt directional reversal could be an issue for them too. We might suppose that animals, understanding this challenge, simply use landmarks and common sense to detect when this 180° shift in their frame of reference occurs, and make the necessary adjustments. Maybe; but for the honey bee, a species that has had millions of years to get this right, the shift can have dramatic, even fatal consequences.

We know a great deal about what's happening in the field because it is possible to train foragers to a food source (typically a sugar-water dispenser on a tripod). A trail of sugary drops from the hive entrance along a walkway to the feeder lures the bees to follow; once a few bees are making regular round-trips on foot, the food can be moved a bit farther away. Eventually the foragers start flying to the feeder. Once they are actually hovering and landing they seem to learn the appearance of the target, and we can move the food away from the hive more rapidly. In general if each forager is allowed to visit once (individual bees are marked with tiny plastic tags or coded paint drops), we can move the feeder about 20% farther away between visits. Thus, at 10 yards the feeder can be moved to 12; at 100 it can jump to 120 yards. It is easy to observe the approach and departure behavior of the bees; if they are making orientation errors, these can be pretty obvious. At the other end of the forager's journey we can observe its dancing and get a direct readout of the bee's opinion about the sun's behavior.

It's clear that an animal that will be using the sun as its compass needs to establish early in its life which direction the sun moves before it can travel very far with any hope of getting back. Once a bee leaves the hive it is quickly out of visual range; if it looks back after even 10 yards the hive entrance will be invisible, and its home tree will look very much like many other nearby trees. At fifty yards it's unlikely that the bee will be able to see that there is a tree at all. For an exploratory flight of, say, 15 minutes near noon in early summer, the bearing back to the hive relative to the sun may have changed by 30° or more. Simply steering along the reverse course from even 100 yards leads to an error of 50 yards—about ten times the limit for finding the entrance visually. Sun compensation is essential; getting the sun's direction wrong doubles the potential error.

It's easy to confirm that bees know the direction of sun movement. Although most foragers choosing to dance do no more than 10 or 20 cycles, there is the occasional marathon dancer. These individuals will perform for minutes or even hours at a time without revisiting the source. Near noon in the early summer when the sun is moving most rapidly, marathon dances visibly precess—that is, in the northern temperate zone the dance angle shifts slowly CCW as the sun outside moves CW relative to the target. The forager compensates for the unseen left-to-right movement of the sun, reporting that the food's relative azimuth is shifting. Marathon dancers are not particularly accurate after the passage of time, at least as the day wears on; presumably their clocks or interval timers are drifting. But under rare circumstances a marathon dancer will start up the next day without having been outside. If the new dance is at about the same time of day as the bee's foraging the previous day, it can be fairly precise. In one extreme case a marathon dancer began again after two months of cold weather, and nevertheless danced accurately. Evidently the ability to entrain their circadian rhythms without going outside is well developed in honey bees. The question of how the compensation is managed over short intervals requires a different experimental approach, as we will see presently.

So if bees can be born north or south of the sun's arc, how do they come to know which direction their primary compass moves? Honey bees typically spend their first two weeks in the hive feeding larvae, tending the queen, and cleaning the hive. As they age they spend more and more time near the entrance taking nectar from returning foragers and guarding the colony from intruders.

During the third week the workers begin taking short practice flights near the entrance. This would be an excellent opportunity to learn what the hive entrance looks like and something about the sun's movement in the sky.

Experiments show that young bees are perfectly willing to believe practically anything, including that the sun does not move at all. Rear them in an indoor flight room with a fixed light and, when trained later outdoors with only celestial cues for guidance, they quickly become lost unless the feeding station moves to keep the same azimuth relative to the sun. But after five days new foragers are tracking the sun. In fact, though, their default assumption seems to be that the sun is stationary. If young bees are allowed out and trained only briefly in the afternoon they seem to assume for at least the first three days that the sun does not move relative to the food. The days bees spend guarding the entrance and making practice flights solve the three-day problem: young bees learn the sun's arc when there is little risk of becoming lost.

The most interesting tests of the sun-direction question were performed by von Frisch's most famous student, Martin Lindauer. In one experiment he set up a hive in Sri Lanka and trained the foragers to a food source about 165 yards to the north. At the time of the experiment in late April the sun appeared to move from right to left in the northern sky. After a week of allowing the bees to collect food near midday he moved the hive north to a location in India where the sun's arc was identical to what they had seen in Sri Lanka, except that now the sun moved from right to left through the southern sky. The few bees that ventured out in the extreme heat of India oriented as though the sun were behaving as it had in Sri Lanka.

This and other tests suggest (but, given the small sample sizes, by no means prove) that bees learn or perhaps *imprint* on the direction of sun movement observed during their stint as guards or while undertaking brief orientation flights, or both. (Imprinting, a powerful force in animal behavior, denotes a special type of automatic learning that takes place during a specific window of time, and is irreversible.) If the sun's direction changes during its two weeks as a forager, a bee that simply learns whether it is in a CW or CCW hemisphere must depend on landmarks until it learns the new solar movement. If unambiguous landmarks are not available to substitute for the sun, the forager will probably lose its way and perish. If, on the other hand, imprinting has occurred (which may well be the case, because there is no clear-cut evidence that any relearning takes place unless the colony is swarming to a new nest site), then the bee is stuck with the wrong information for the last few days of its life.

For vertebrates, there has been less interest in the sun-direction question. These relatively long-lived species have to be able to recalibrate themselves to what must, whether in the home range or during migration, be a dramatic change when their migrations take them past the latitude at which the sun's arc passes through the zenith.

■ How Fast?

The third problem is compensating for the changing rate of sun-azimuth movement. Fifty years ago, when animals were thought of as simpleminded robots, most researchers assumed that the solution to the problem must be for the animal to use an average rate of azimuth movement—15°/hour—and just live (or die) with the errors. But the work of von Frisch and others made abundantly clear that honey bees are much more precise than this. As we saw in the previous chapter, departing foragers imprisoned for an hour nevertheless correct for the sun's unseen movement with good accuracy. The first behavioral strategy to suggest itself was that animals somehow "know" the sun's arc through the sky. Perhaps they are actually computing the answer. For humans this would require tables and a calculator, but a hardwired spherical geometry module could solve the problem graphically. This solution, however, requires knowing (or compensating for) latitude and date.

In theory, simply remembering the previous day's angle between the plane of the rising sun's arc and the horizon, plus the maximum elevation at noon, would allow an animal to reconstruct the sun's path through the sky. The animal could subconsciously move the sun along that arc at 15°/hour, drop a line from each position vertically to the horizon, and compute the corresponding azimuths. An alternative to this spherical-geometry solution would be to memorize the azimuth for each time of day, creating a mental table, or to learn each combination of azimuth and elevation as a function of time.

But while researchers have focused on strategies that mimic human navigation, the evidence from animals is clearly against any approach that involves knowing elevation or date. Clock-shift tests are particularly revealing. Researchers place an animal in an isolated room with an artificial dawn out of sync with clock time; in this way they can move an animal to a new time zone or reset its internal clock over a short period. Animals with a time–azimuth–elevation table that were subjected to a clock shift would not be able to reconcile the three columns in memory. For the internal time the animal registers, the observed elevation would be wildly wrong; thus, it would not be able to use the apparent sun azimuth to orient its flight. Orientation tests show that birds and bees believe their clocks and relate this time value to azimuth; they ignore solar elevation in their compass behavior. What are they doing, then?

To look at the question quantitatively, we trained bees from a hive to a feeding station in a large field. These tests were carried out in late June near the summer solstice, when the sun's movement in the temperate zone is more variable than at any other time of year. The observation hive was closed for an hour either at 11 a.m. (solar time) or at noon. This confinement is not uncommon;

a summer shower or a period of high wind (anything above 15 mph) will do the same. The hive was moved to another field to limit the use of landmarks; foragers had to rely on their guess about the relative azimuth between the sun and the food after an hour's absence. When the foragers reemerged they found an array of stations available—an array that included the same compass direction as during training, the direction a crude 15°/hour correction would specify, and (inadvertently in the first test, but on purpose after that) the bearing that corresponded to an extrapolation of the sun's rate of azimuth movement just before the hive was closed. We recorded the arrival directions.

The results indicate that without landmarks the bees are using the past to judge the future, extrapolating the sun's azimuth change. Under both conditions, however, the extrapolation that the bees display is somewhat smaller than the actual value we calculated. Could the extrapolation be out of date or, which is the same thing, be based on an average over several previous measurements? To look at this we trained the bees to a very prominent feeder with a large flag visible from a considerable distance and recorded the dances. From time to time the feeder was moved to a different azimuth—from 220° to 250°, and then later back to 220°. This created no problem for the foragers, because both the feeder and the large barn housing the hive were visible from a distance. The bees steered directly for the target. Their dances, however, were initially unaltered after each shift, and only slowly drifted back to the correct relative azimuth. Newly recruited bees, on the other hand, danced accurately upon their return.

The foragers were registering the sun's position but averaging it with previous measurements; only after about 40 minutes did the effect of the shift fully disappear. This probably means that foragers average across the last 40 minutes, a tactic that reduces measurement errors by virtue of combining several estimates, while minimizing the effect of changing azimuth. Looking back at the hive-closing tests, a 40-minute sampling period would account for the

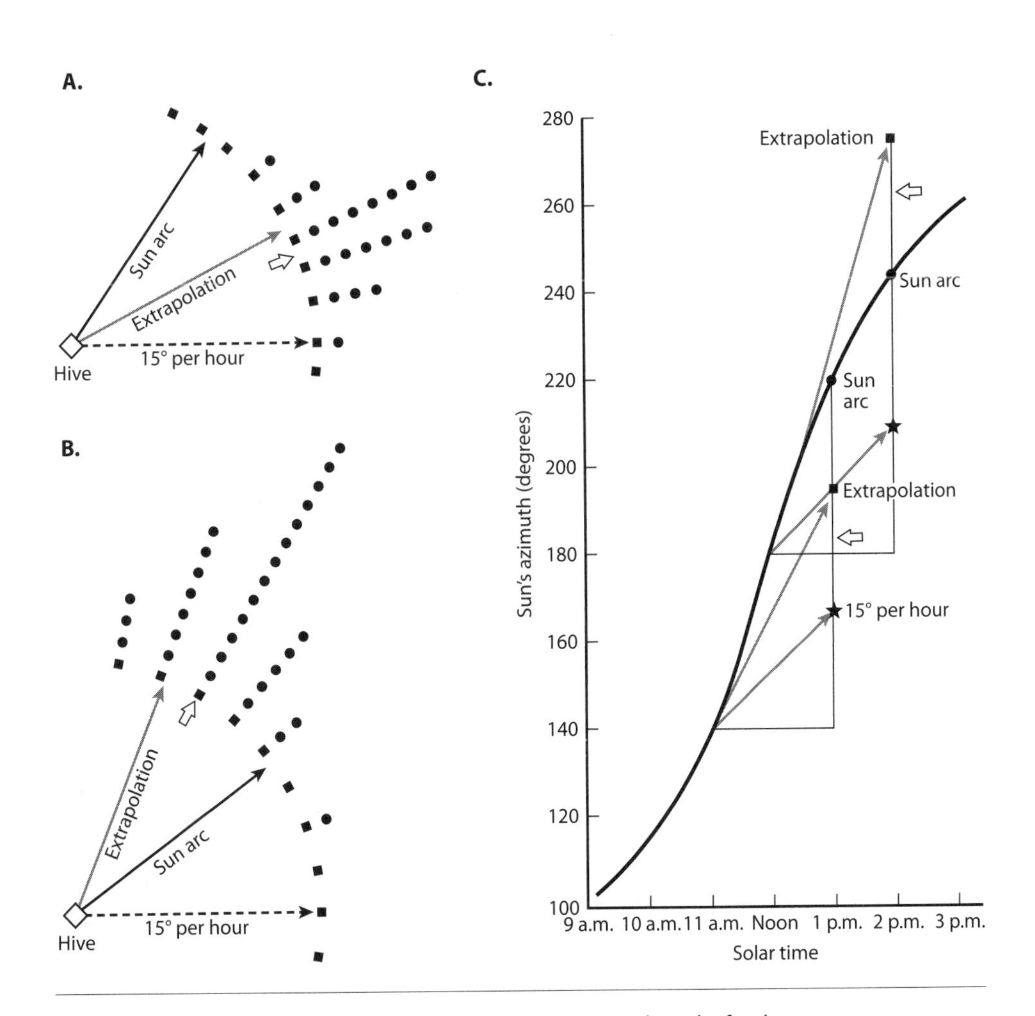

Judging the sun's movement. (A) The testing array and results for the 11 a.m.–closure experiment. Each capture station in the array is shown as a square. The circles represent the direction of arrival of trained foragers. The white arrow is the mean direction. (B) The testing array and result for the noon-closure test. (C) The change in sun azimuth is shown by the curve. For both 11 a.m. and noon the predicted arrival azimuth of foragers is shown based on whether they use a 15°/hour approximation, know the sun's arc, or extrapolate the sun's movement.

underextrapolation observed there. The movement-rate values the bees appeared to be using were about 20 minutes out of date, which is just what you get averaging over the previous 40 minutes. We will look presently at the remarkable elaboration of this strategy that bees use when they have landmarks available.

Our human diurnal bias is at odds with the realities of migration, which takes place almost entirely at night. A long-distance traveler must worry about predators and the danger of thermal stress: no behavior produces more heat than flying, and doing so with the sun beating down requires large quantities of water to keep from cooking the flight muscles. Thus it should be no surprise

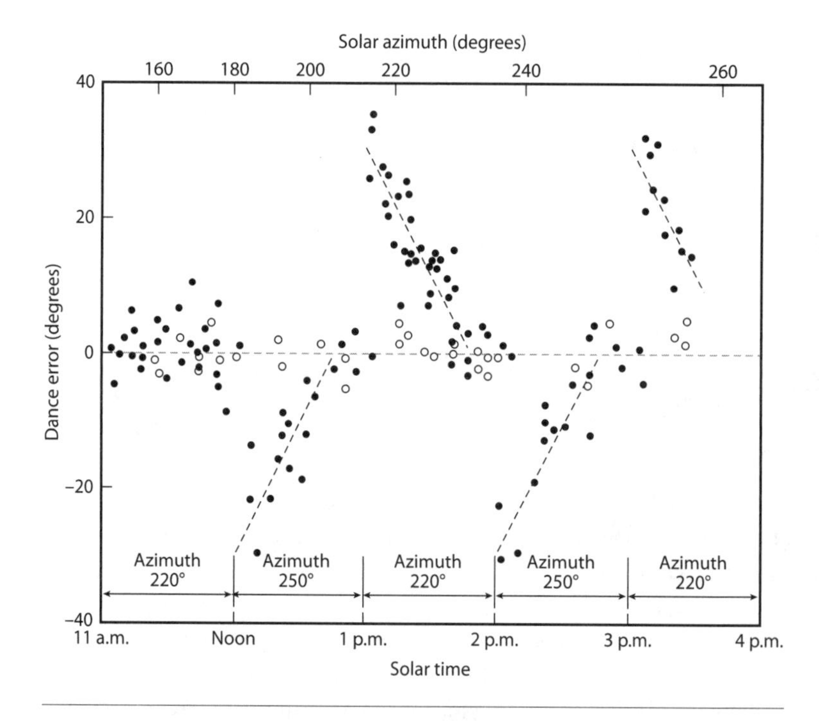

Integration time. When the azimuth of a highly visible feeder is shifted by 30° the dances performed by returning foragers (closed circles) only slowly adjust. The dances of new recruits (open circles) are accurate from the outset. Apparently bees integrate sun-azimuth data over the most recent 40 minutes; as a result, the average is about 20 minutes out of date.

that most birds and many insects travel after dark. But what do they do without the sun?

One potential alternative is the moon, but remarkably little work has been done on the possibility of a lunar compass. In part this is because birds, with their high visual acuity, can use the far more regular motions of the stars. And, of course, the stars are visible on any clear night, whereas the moon is above the horizon only half the time. As we will see, the earth's magnetic field provides a relatively simple compass, free of the need for time compensation and landmark calibration.

The sandhoppers that provide such a good example of innate orientation also have been studied intensely for what they can tell us about lunar compasses. During the day they use the sun as their guide, but at night substitute the moon if it is visible. Confinement tests, in which they are allowed a view of the moon and then kept in isolation for varying amounts of time before testing, are revealing. The moon's azimuth in one case moved 103° over about 6.25 hours. The sandhoppers' orientation accommodated all but 18° of that shift—plenty of accuracy for the task in hand. The pattern of errors is intriguing: the sandhoppers' estimate of the moon's azimuth lags behind its actual movement, especially when it is rapid. It is as though they are using the average value of 15°/hour rather than learning or computing the arc.

■ Polarization

Diurnal navigators seem to concentrate on the sun as the primary orientational cue. They recognize it without being able to resolve its disk, and compensate for or even anticipate its movement. But a simple modification of the original test on ants demonstrating sun-compass orientation reveals a serious complication in this neat solar navigation picture. Rather than providing a mirror image of the sun, try blocking a direct view of it instead. If ants had a simple sun-based navigation system, blocking the sun instead of providing a mirror image should cause them to wander randomly. In fact, provided the ant has a generous view of a clear sky, it continues along its route unfazed. Similarly, a forager bee dancing on a horizontal surface in the open has no problem orienting her dances even if we interpose a barrier that obscures the sun. Clearly there is a second system that allows insects to infer the sun's location from information in the rest of the heavens. Karl von Frisch was able to show that bees need only a 10° patch of clear sky, featureless to humans, to orient their dances.

Von Frisch guessed that this backup system might be based on patterns of polarized light overhead. Humans are blind to this cue, but you can readily demonstrate to yourself that blue sky is highly polarized if you have the right filter. Take a pair of polarized sunglasses and hold them up at arm's length roughly 90° away from the sun; rotate them and you will see the lenses go black and then, a quarter turn later, they will seem transparent again. If you move the glasses around a bit, you should be able to detect a darker band in the sky oriented at a right angle to the sun.

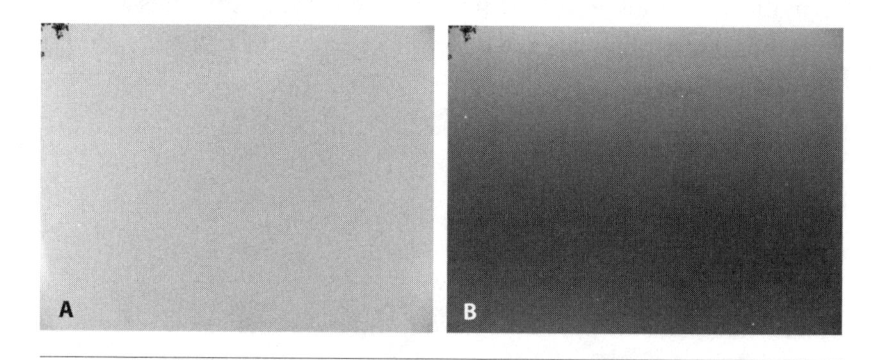

Polarized skylight. When holding a polarizing filter between your eye and blue sky 90° away from the sun, at one angle the sky will appear bright, whereas a quarter turn of the polarizer will turn the view nearly black. These pictures were taken in the afternoon looking NE, with the sun in the SW (out of the picture above the image).

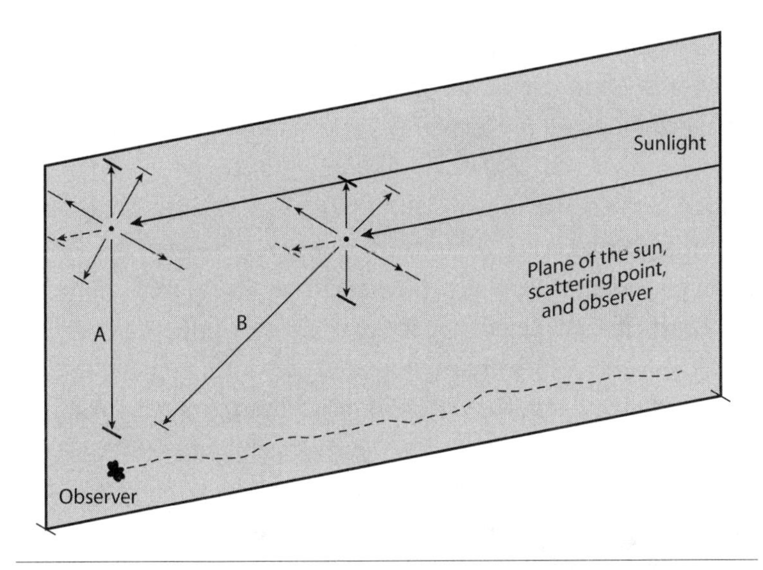

Scattering in blue sky. Light from the sun is unpolarized. After scattering in the atmosphere some of the light becomes polarized perpendicularly to its plane of travel. It is most strongly polarized when it scatters 90°, and less strongly as the angle approaches 0° or 180°.

Polarization arises when sunlight scatters off of air molecules. Each photon of light is a wave vibrating at a particular angle as it travels. Light from the sun is unpolarized and white—which is to say, it is a mixture of all angles of polarization and all colors of light. Each photon races through the atmosphere until it interacts with an atom. The probability of such an interaction is related to wavelength: short-wavelength light—ultraviolet and blue—is far more likely to scatter than longer wavelengths. As we noted above, this is why the sky is blue. (It also explains why the sun is orange at the horizon: most of the blue and green are scattered out of the light's long, oblique passage through the atmosphere, leaving behind red, orange, and yellow.) The scattering interaction in the sky overhead imparts a polarization to the outgoing light; the photons of light scattered to earth are oriented—polarized—in directions that produce the effect of dark and light in the sky to animals equipped to detect it. Photons scattered 90° are highly likely to be polarized, those scattered 0° and 180° (ahead and behind the incoming path) are unpolarized, and those at intermediate angles have a probability of becoming polarized related to the scattering angle.

Primary scattering creates a dramatic pattern overhead. For an observer looking at the sky, every point on the celestial dome lies on a particular plane passing through the sun and the navigator's eye. Two such planes are shown here, each tracing a circular arc overhead. The light at every point along the arc is polarized perpendicularly to the plane. Moreover, all such planes intersect the sun, providing a theoretical way to localize it even if the sun is behind a cloud or below the horizon. Bees, in fact, have added a processing trick that allows them to guess the sun's azimuth without performing spherical geometry. How birds and other vertebrates with their high-resolution eyes decode the pattern simply is not known.

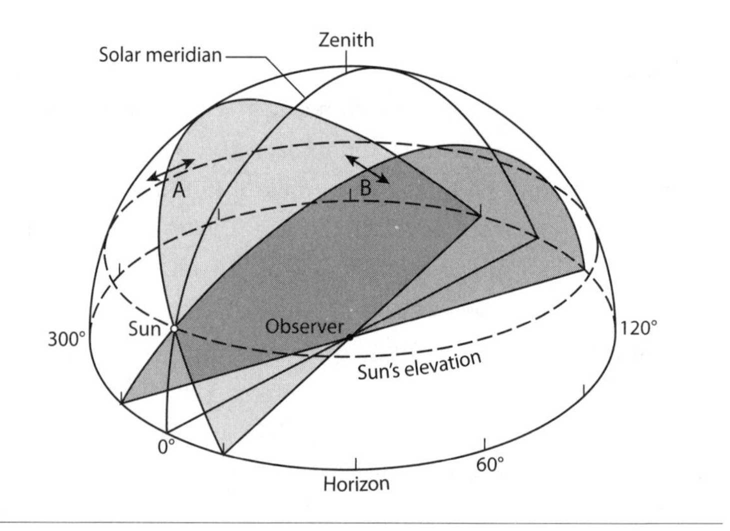

Two scattering planes. Because scattered light is polarized perpendicularly to its plane of travel, each plane has a unique angle of polarization.

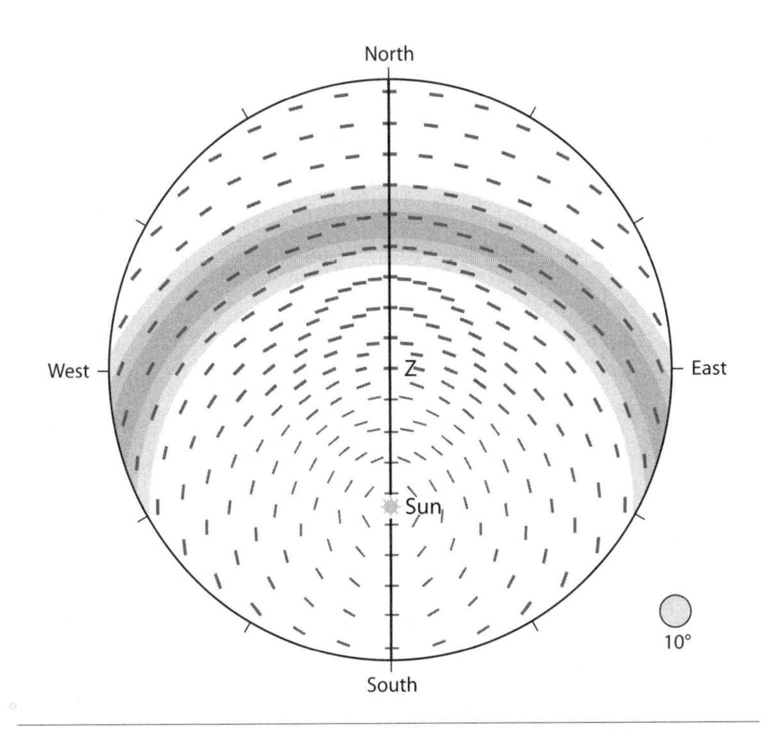

Pattern of polarized skylight. Allowing for the difficulty of representing a hemisphere on a flat piece of paper, at any point overhead the pattern of polarized light in a clear sky is perpendicular to the plane also containing the sun and the observer. The circular bands of intensity (indicated here by the thickness of the polarization lines) reflect the fact that the degree of polarization depends on the angle of scatter, being strongest at 90°. This illustration represents the sky at noon on the equinox at latitude 42°. The small circle in the lower right represents the minimum 10° circle of sky needed by honey bees to orient themselves.

We've focused on clear-sky polarized sunlight. But given the ability of some insects to see by starlight, there is a far-fetched possibility that the light of the moon might create a similar ultraviolet pattern in the night sky. The first insects to demonstrate the reality of this theoretical detection system were dung beetles, which scour the ground at night for animal scat. They shape it into balls, excavate a burrow, lay an egg on each dung ball, and bury them. Despite the irregular features of the dung balls and the uneven ground, their path away from the dung pile to the burrow is remarkably straight and true. No wandering in circles for these insects.

If researchers interpose a polarizing filter between the beetle and the moonlit night sky, being careful to block a view of the moon itself or choosing times when the moon is just below the horizon, the beetles adjust the direction of their routes. When the filters are rotated the beetles reorient by about the same angle. In fact, they will often ignore the moon even if it is directly visible and concentrate on the polarized light. Even very dim polarization, then, can be a cue for nocturnal insects, though in this case there is no reason to believe it is time compensated or used to derive the direction of the light source.

The ability of the beetles to still read the pattern when the moon is below the horizon reminds us that two of the best times for observing polarized light patterns are just before dawn and just after dusk. The sun is below the horizon but the pattern of scattering is clearly visible in the sky overhead. The band of maximum polarization will typically pass through or near the zenith. For crepuscular species—those most active near dawn and dusk—the obvious cue to use is polarization, because the sun and stars are invisible. Indeed, the Vikings discovered this possibility while navigating during the long twilights so common in the far north. By looking at the reflection of the sky off a calcite crystal and scanning the horizon for the darkest image, they could pinpoint the azimuth of maximum polarization; the sun was inevitably at the brighter of the two spots 90° away.

■ Clouds and Landmarks

Navigation by piloting employs a learned map of local landmarks. We will look in a later chapter at how animals do this. But landmarks are also used during sun-compass orientation as a way of sorting out ambiguities—including the one created by the polarized-light rule. Though in general a bee at any given moment is using the sun or polarized light or a remembered landmark, it seems to be keeping track of them all just in case. Probably the most dramatic and eerie illustration of this is seen in the honey bee's dance on a cloudy day. Bees fly to and from food sources under overcast, and perform well-oriented dances in the hive, dances keyed to the azimuth of a sun they cannot possibly see. How is this possible?

The crucial test took advantage of a phenomenon von Frisch discovered that we like to call landmark overload. If you train a group of foragers to a good source of food along a prominent landmark (NE along a forest edge running SW–NE, say, so the outward-bound bees fly NE with the trees to the right), the foragers become very attached to the visual cues en route. If you then move the hive overnight to a new location east of a forest edge running north–south, the vast majority of trained bees will fly out south along the forest, keeping the trees again on their right; only a few will choose the correct compass direction (NE). The landmarks have gained more salience than the sun.

Our lab repeated this relocation, but tried it on both sunny and cloudy days. We looked at where the foragers went, which direction they danced, and at which alternative feeding station recruits arrived. As expected, nearly all of the foragers in the displaced hive cued off the trees under both sky conditions, flying to the landmark station. But the dancing in the hive was dramatically different. On sunny days the relocated bees patronizing the feeder along the trees aimed their dances south, indicating the true (and new) celestial direction of the landmark station; the few traveling to the compass station aimed their dances at it. But on cloudy days bees flying to the landmark station performed dances indicating the old direction: NE, the location of the compass station. Lacking any compass in the sky, no foragers found the compass station under

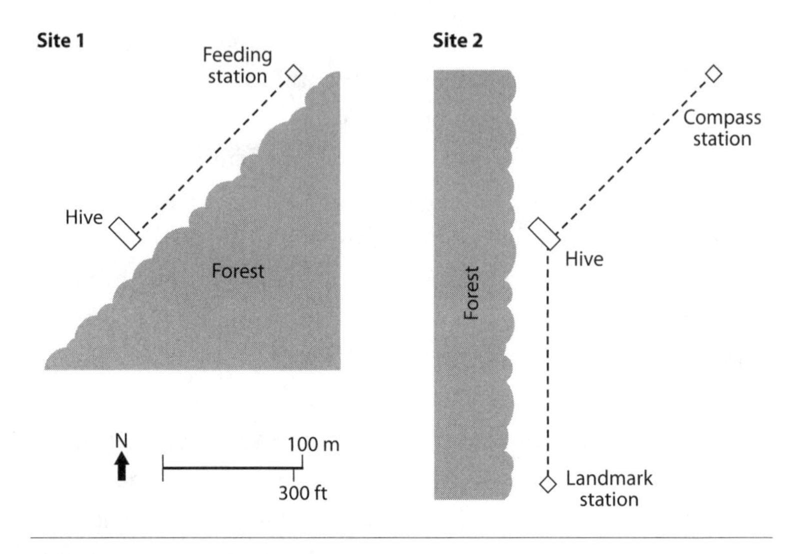

Experimental layout in the hive-movement tests. Foragers are trained along the forest edge to the NE with the trees to the right on the way out. After a few days of visiting the feeding station the hive is moved overnight to a forest edge running north–south. Most foragers again fly out with the trees to the right, which takes them south to the landmark station rather than NE to the compass station—the feeder in the same training direction.

overcast. The bees were simply remembering the dance direction from the previous day. Moreover, recruits attending the cloudy-day dances pointing NE dutifully flew along the trees, in the direction that had been NE previously. They were apparently using their memory of the sun's azimuth at that particular time of day relative to the forest edge.

Memorizing solar azimuth with respect to landmarks or a compass cue (such as magnetic north) for each time of day, and updating this matrix whenever possible, seems to be a core strategy of both insects and birds. It could be that bees remember the path of the sun relative to local landmarks, allowing them to substitute a likely value when neither the sun nor blue sky is visible. Indeed, later tests showed that bees trained to new food sources use their time-linked memory of the sun's azimuth relative to landmarks to orient new dances. On sunny days this memory might resolve the occasional directional ambiguity that arises from using a restricted patch of polarized light. The bees' ability to extrapolate the sun's rate of azimuth movement might be involved.

But a puzzling observation suggests there must be even more in the insect's celestial navigation repertoire. Bees allowed out of their hive only in the afternoon can, after 5 days, orient correctly to the morning sun. That is, they can compensate for the sun's moving azimuth without having experienced it at that time of day. Conceivably this could involve some sort of reverse extrapolation; the pattern of movement after noon is a perfect mirror image of what the sun does in the morning. With time-based memory and

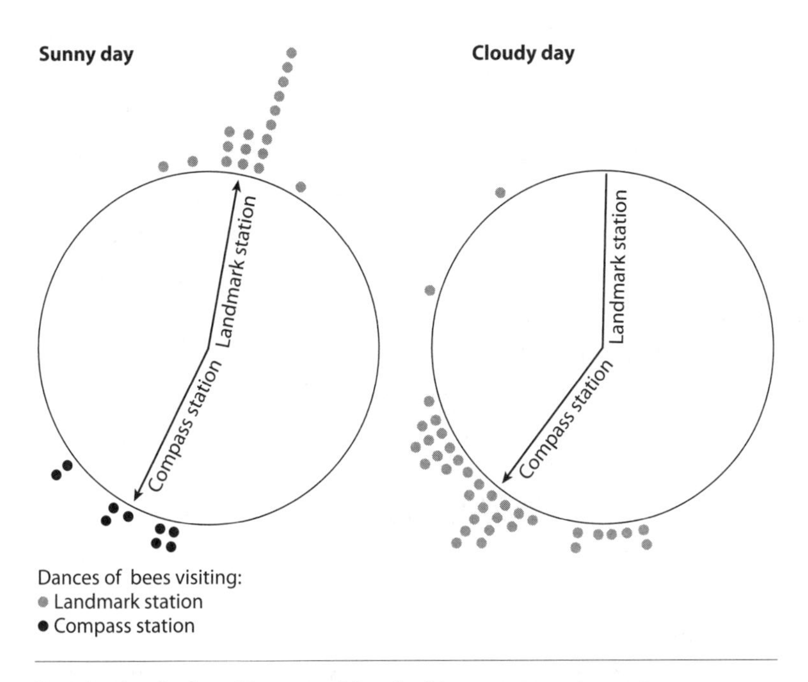

Dancing by displaced foragers. After the hive was moved most foragers followed the trees to the landmark station on sunny days, and all did under overcast. With the sun visible the dances indicated the site actually flown to; under overcast, the dancing was oriented toward the unvisited compass station.

landmark-cued calibration of the sun's behavior, bees might be able to fill in the missing pieces. But the actual processing trick remains a mystery.

■ Can Invertebrates Sense the Earth's Magnetic Field?

Except in imaginative writing, scientific study of magnetic field sensitivity is relatively recent. A hundred years ago "magnetic intuition" was a catchall hypothesis for practically anything unusual, including the bees' ability to communicate about flower location. Some weak results that proved difficult to reproduce were reported for birds around 1950, but even researchers who took the observations seriously had to admit that the behavioral response actually measured was much too weak to be of any use in navigation. The early 1960s saw some puzzling but consistent results with termites, flies, and snails. With all other cues blocked, the animals took up specific orientations relative to the earth's magnetic field when allowed to stand or move spontaneously. The angles involved, however, had no obvious relevance. On the other hand, this sort of "nonsense orientation" is observed in many species at rest under polarized light or in the presence of other cues or gradients that are, in at least some situations, used sensibly. The earth's magnetic field has lots of potentially useful information for local navigators and long-distance migrants alike. Showing that animals could actually make clear and adaptive use of it was the problem.

In the late 1960s the honey bee settled the matter for invertebrates, though the discovery raised many more issues for researchers to investigate. German scientists first discovered that a small but systematic error of a few degrees in the direction indicated in the dance language disappears when the magnetic field felt in the hive is canceled—that is, when the hive is surrounded by a set of precisely built coils and enough current is sent through the wires to exactly compensate for the earth's field. The residual misdirec-

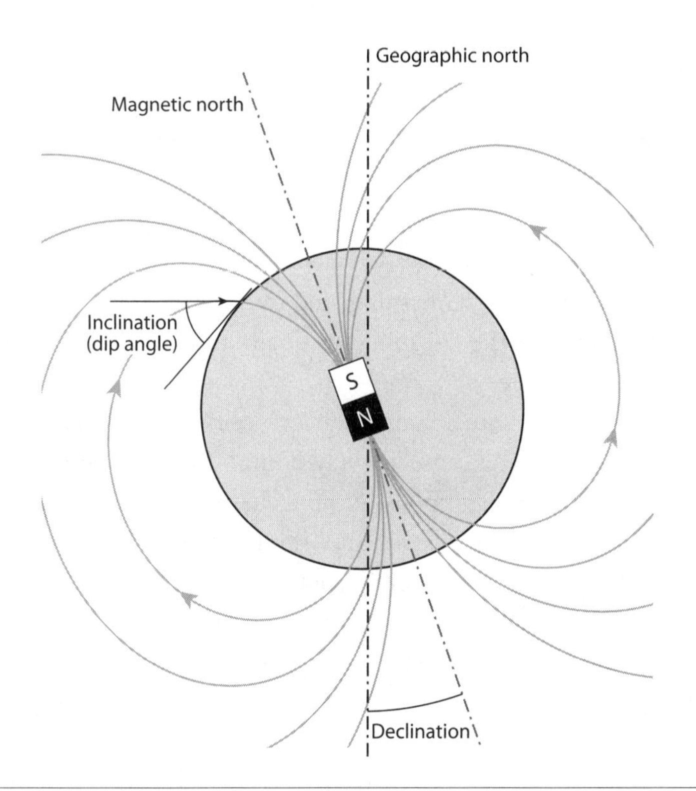

Magnetic field lines encompassing the earth. To a rough approximation, the earth's core acts like a weak bar magnet 1000 miles long buried 4000 miles below the surface and tipped 11° from the planet's axis of rotation—that is, the line connecting the geographic south and north poles. This discrepancy gives rise to the declination. The lines of force curve from one pole to another, leaving the earth at an angle (the inclination, or "dip angle") that depends largely on latitude.

tion in the dance is probably an artifact of magnetic sensitivity. Because the recruits and foragers make compensating errors, the communication system functions elegantly.

A second intriguing result involved a kind of nonsense orientation. During our lab's horizontal hive tests we sometimes turn off the experimental sun and the patch of synthetic sky we have been showing the bees to get some control data. To our surprise, the disoriented dances that resulted had a pattern, with four directions preferred. After puzzling over this we realized that they were oriented toward magnetic north, south, east, and west. The feeding station was SW at the time, and the behavior did not change when we trained in other directions. When we canceled the magnetic field the pattern disappeared. Martin Lindauer's group in Germany discovered the same thing, and found that when the earth's field is amplified the inexplicable quadramodal preference becomes even stronger.

Given that bees are unambiguously sensitive to the earth's field, what is the magnetic sense actually doing? There are at least two important uses, both responses to the challenge of living inside a dark cavity. The first is that all-important need to calibrate the circadian clock. For an animal living outdoors, observing dawn is not a great challenge; but for the thousands of worker bees needing to reset their clocks before setting out to forage, sunrise would pass unnoticed.

While the earth's magnetic field is mostly generated by the core, a small part arises inductively from ions carried west to east in jet

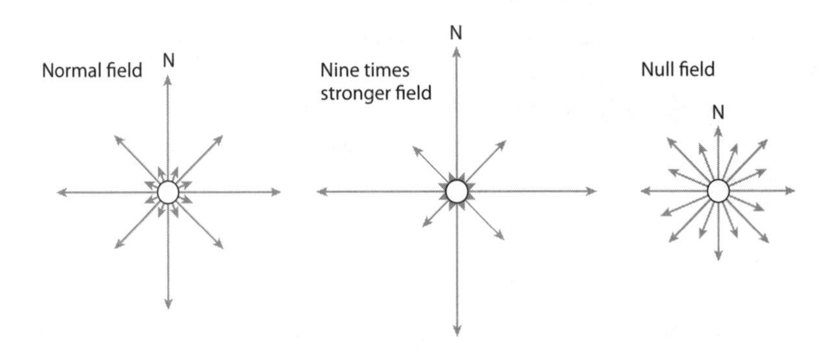

Nonsense orientation in bees. In a horizontal hive with no visual cues dancers orient to the eight points of the magnetic compass—most strongly to north, south, east, and west. When the magnetic field is strengthened the preference for the four major directions increases; when the magnetic field is canceled the behavior vanishes.

streams in the upper atmosphere. (We will explain how induction works in the next section). Because the earth heats and cools in a regular cycle thanks to the warming effect of sunlight, the atmosphere in the tropics expands slightly in the day and contracts a bit at night, pushing the jet streams alternatively north and then pulling them back south. The field induced by the ion flow changes the field strength felt on the surface by about 0.4% up and down over the course of a typical day. Bees in a hive isolated in a basement permeable to magnetic fields—one built without steel, and free of iron reinforcing rods in the concrete—can maintain a circadian rhythm indefinitely. Add a small but randomly varying magnetic field and they lose all sense of time. This behavior requires a degree of sensitivity to the field far, far beyond what is necessary for a mere compass. It was the first hint that led eventually to one possible solution of that mystery of mysteries, the map sense of animals.

Another way bees use the earth's field affects their annual swarming. Each spring a healthy, growing colony will send out a swarm of up to 20,000 workers, plus the old queen. (The rest of the workers and a new queen remain in the natal cavity.) The swarm forms a cluster in a nearby tree and dispatches scouts to look for a suitable cavity in which to build a new home. The scouts report their finds by dancing on the side of the swarm itself, on a stage of living bees. They visit and compare each other's finds. Once a consensus is reached they fly as a group to the new cavity.

The first priority is to build new comb so that eggs can be laid, larvae reared, and food collected and stored—a race against time since the colony must accumulate ample stores to survive the approaching winter. The completed hive will have several parallel sheets of comb, each two bee-diameters from the next. But with hundreds of bees engaged in building these intricate arrays of cells in compete darkness, what keeps the construction accurate and orderly, rather than random?

Every natural cavity is different in shape and size. If the space is elliptical the bees build along the long axis; other shapes result in

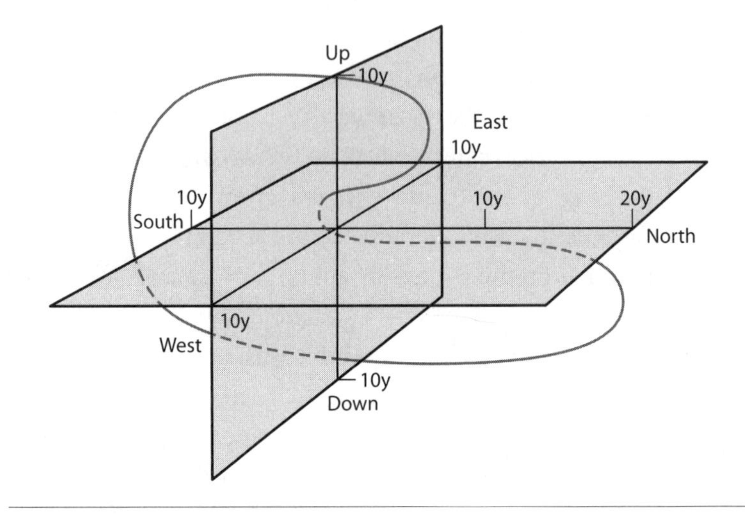

Magnetic clock of bees. The earth's magnetic field strength at 45° latitude is about 50,000 γ. The daily variation (shown here in all three axes over the course of a day, measured in Europe) is typically on the order of 40 γ. The pattern is different at other locations but relatively consistent from one day to the next. Magnetic storms, produced ultimately by sunspot activity (which injects enormous numbers of new ions into the atmosphere) disrupt this pattern—and the ability of bees to stay synchronized with dawn.

different configurations. But even if there is no clear best solution, the workers agree: each bit of wax added in the darkness of the hive has a consistent orientation. In a perfectly uniform circular experimental hive with the entrance in the center, new comb is built at exactly the same angle to the earth's magnetic field as in the old hive. Rotate the field, and new comb is built accordingly. Put a strong magnet on top of the hive creating a radiating field, and the bees will fabricate rings of comb. When there is no obvious directional logic, then, bees draw on their magnetic sense to replicate the orientation of their previous colony.

Subsequent tests have shown that bees can be trained to distinguish between two alternative feeders based on the local magnetic field direction or strength, another impressive ability but one of dubious practical value. The obvious question is whether bees use their magnetic compasses for navigation when the sky is overcast, or when there is a polarization ambiguity to resolve. The simplest way to test this is to glue tiny magnets onto the foragers and see what it does to their orientation. Alas, the very things that make honey bees perfect for so many things frustrate this approach. The foragers all return home, enter the hive—and their magnets stick together.

A magnetic compass sounds like a great thing to have, a perfect fixed-direction system free of time or the sun's fickle movements in the sky. The reality is less wonderful. The first problem is that this compass rarely points north. The earth is a sloppy magnet; its field arises from the rotation of the planet's molten iron core, in which complex eddies constantly form and shift. The actual magnetic poles are hundreds of miles from the geographic poles, and wander many miles each year. The distorted field lines produced by this semispherical magnet are not nearly as regular as the neat pattern associated with a conventional bar magnet. Moreover, the strength and direction of the field at any point on the earth also are affected by the rocks underneath; particularly in iron-rich soil, the effect can be on the order of 10%.

All this means that animals cannot simply consult a magnetic compass and read "north." Instead they must compensate for the *declination*—the specific deviation from true north characteristic of their location. This calibration inevitably involves the sun or stars. Added to this extra step is the apparent noise in the measurements; behavioral tests indicate that despite the problem of time compensation, animals can read their celestial compasses with better accuracy than their magnetic ones.

Evidence that a variety of other invertebrates use a magnetic compass for navigation ranges from suggestive to clear-cut. Monarch butterflies, for instance, continue to fly toward Mexico in the autumn under overcast conditions; a strong magnetic pulse at

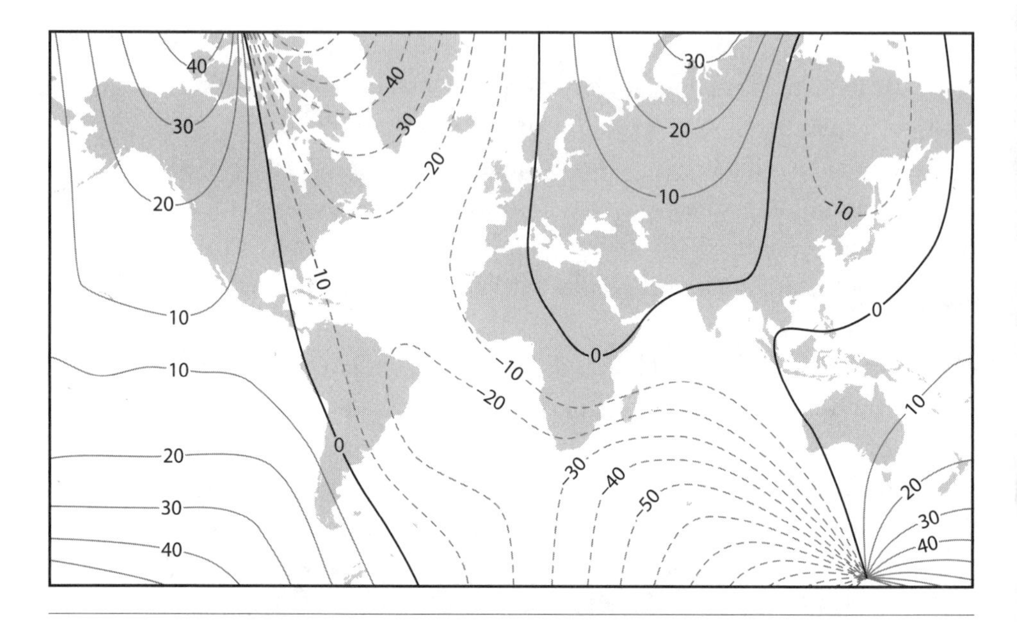

Magnetic declination. Because the magnetic poles are located at some distance from the geographic poles, and because the earth is a poorly organized magnet, a compass needle has a declination (a deviation from true north) at almost all points on the earth. The error is about 20° in Nova Scotia, eastern Brazil, and South Africa, for instance. Most species need to calibrate their magnetic compasses, and long-distance navigators need to recalibrate as they travel out of their home area.

least temporarily destroys their ability to choose the correct direction on cloudy days. Nocturnally migrating moths are well oriented with no celestial cues; a magnetic compass is the likely explanation for their ability. The spiny lobster, a species that migrates over about 30 miles in tandem conga lines antennae to tail, maintains its consistent orientation even at night and across the irregularities of the ocean bottom. Their ability is clearly magnetic: most spiny lobsters refuse to be trained, but about 40% of individuals are willing to learn that food is in a consistent direction. Rotate the field and the trained lobsters will march in the new magnetic orientation.

■ Sensing the Field

There has always been something of mystery and romance associated with magnetism. It is an invisible force that attracts or repels, and rotates compass needles into alignment with the magnetic poles (or the Holy Land, as Europeans initially thought). Understanding the basis of magnetism hardly reduces the mystery, though it makes clear how an animal might sense the field.

The basic phenomenon behind magnetism is actually electromagnetism, the movement of electrons that creates a surrounding magnetic field. The electrons themselves are tiny magnetic dynamos; each has a spin, and that rotation creates a miniature field complete with north and south poles. In isolation an electron will align its private magnetic field with the earth's field, a phenomenon known as *paramagnetism*. This adds minutely to the earth's field, and thus paramagnetism magnifies the local field. More important for most electrons, however, are the fields of its nearest neighbors. Because magnetic strength falls off exponentially with distance, proximity can be hugely important. A pair of adjacent electrons (four in these parenthetical illustrations) have two stable orientations: spins aligned end to end (→→→→), or spins oppositely directed side to side (↑↓↑↓). As we all know from experience, magnets will stick together side to side with the north pole of one next to the south pole of the other; they will also attach end to end, with the north tip of one attaching itself to the south tip of the other. For electrons it's the same thing at the subatomic level. Side-by-side pairs are more likely to form spontaneously, and their fields cancel one another. End-to-end arrangements last only if they are stabilized by other well-aligned electrons, but in this case the fields add to one another.

In certain crystals the spacing and positioning of the atoms causes these spins to line up and reinforce one another, creating a permanent magnet. The iron oxide mineral magnetite ($FeO \cdot Fe_2O_3$, also known as lodestone) is the most famous example of this *fer-*

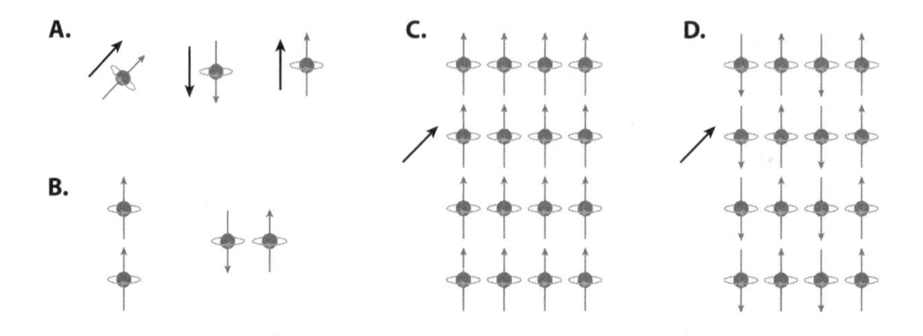

Electron spins. (A) Isolated electrons will align their spins with any outside field—most often the earth's (the black arrow to the left); this is simple paramagnetism. (B) Pairs of electrons close enough together will generally align their fields either end to end or side to side; the latter is more stable. This is a superparamagnetic interaction. (C) In certain special substances the end-to-end spacing of atoms with unpaired electrons can dominate over the side-to-side spacing, producing (if the crystal is large enough) a permanent magnet—ferromagnetism. (D) In most crystals with unpaired electrons, the side-to-side interactions dominate, and the local fields cancel one another.

romagnetism. But even ordinary iron can hold at least a temporary field, as when a needle is stroked in a consistent way with a permanent magnet.

As a magnetic crystal grows it behaves paramagnetically at first as its unpaired electrons track the external field of the earth. As we've said, because of atomic spacing the fields of these odd electrons can reinforce one another, creating a magnet temporarily aligned with (and locally amplifying) the earth's field. Such interacting assemblies are said to be *superparamagnetic* to distinguish them from the ordinary paramagnetism of single atoms or molecules. Once the crystal is large enough its internal fields are stronger than that of the earth and it becomes a permanent magnet, typically taking on the last alignment it felt before it achieved stability. This is the phenomenon that allows geologists to track the movement of continents, because the last field felt by a growing or cooling crystal is fossilized in the rock.

Serious human navigators are all too familiar with the distinction between superparamagnetism and permanent magnetism. In theory, using a magnetic compass should be trivial: hold it steady, observe the direction of the needle, apply the correction for your location to convert magnetic north to true north, and you are set. As we have seen, though, local magnetic effects can overwhelm the earth's field. Many early sailors refused to use this tool because it often literally steered them wrong. Of course, early ships were full of iron, from nails and barrel hoops to stair rails and cannons. Any ferrous object near the compass could influence the reading, but in maddeningly different and unpredictable ways. One problem was "hard" iron—iron with a small but permanent magnetic field. This ferromagnetism was generally induced during fabrication, particularly as a piece of wrought iron cooled. At least such fields are constant; a simple correction or carefully placed compensation magnets could cancel the problem. More insidious was soft iron—iron in which the earth's field induces superparamagnetic fields to which the compass needle then responds. Though a superparamagnetic field is parallel to the earth's field when very close to the domain in question, at increasing distances the direction changes as the field lines loop back from one pole of the grain to the other—or to other domains or objects. Even if the problematic piece of metal was fixed to the deck, the induced field would shift as the ship turned relative to the earth. Compensating for soft iron effects is extremely difficult, but the failure to do so can be fatal; the history of how this problem was eventually solved is a classic mixture of insight, luck, and perseverance. Animals, on the other hand, have evolved to *use* paramagnetism rather than worrying about how to cancel it out.

In designing a magnetic field detector to sense the earth's field, insects have the three options just mentioned: permanent magnetism, superparamagnetism, and paramagnetism. Insects are too small for a fourth strategy, *induction*, which involves moving a conductive loop through the field. For centuries the possibility of magnetic sensitivity has been controversial at best, and more often the object of academic scorn. It comes as a humbling surprise to know that numerous species from bacteria to burrowing mammals monitor the earth's magnetic field, and all four theoretical possibilities have evolved independently.

In animals the first possibility, paramagnetism, involves using the energy of light. This energy can create a pair of highly reactive molecules in the retina, each with an unpaired electron. The spins of these two electrons will be parallel (↓↓) or opposite (↑↓), depending on the local magnetic field. Just how long this activated form lasts before the molecule relaxes back to its ground state also can help encode the strength and direction of the local field.

Testing this hypothesis seems trivial. Ask a magnetically sensitive animal to orient in darkness; if it fails, a light-dependent mechanism is involved. If the strategy really does involve photons and paramagnetism, only one wavelength should work. Some animals, such as the fruit fly *Drosophila*, cannot in fact orient magnetically in the dark. When the test is repeated under dim monochromatic light the ability returns, but only with blue and green wavelengths. Painting over the eyes demonstrates that the receptor is part of the visual system.

The pigment in question is *cryptochrome*, a blue-green–absorbing molecule not involved in insect vision but that rather is a light detector used for a variety of tasks in different organisms. In bacteria it responds to mutagenic light and activates the DNA-repair system. In plants it has a role in controlling circadian rhythms; in corals the molecule is part of the lunar-cycle detector. Flies without the gene for cryptochrome cannot orient to magnetic fields. Cryptochromes and their paramagnetic operation create a detector with a very odd property: animals using this approach cannot directly distinguish north from south. Cryptochrome-based compasses instead use the *dip angle* or *inclination* (the angle between a line of force and the earth as the line curves from pole to pole) of the magnetic field to judge north. Such compasses correctly assume that the nearest pole is in the direction of downward field slant.

Superparamagnetism is the most likely mechanism behind the extreme sensitivity of honey bees. Along with Princeton colleague Joe Kirschvink we originally looked for magnetite in bees using a superconducting magnetometer normally reserved for geologists tracking the orientation of rock samples from the earth's crust. We discovered almost immediately that all surgery must be done with plastic or glass; steel scalpels and razor blades inevitably shed microscopic bits of hard iron that obscure the weak fields of anything biological. And thorough washing is crucial, because dust from

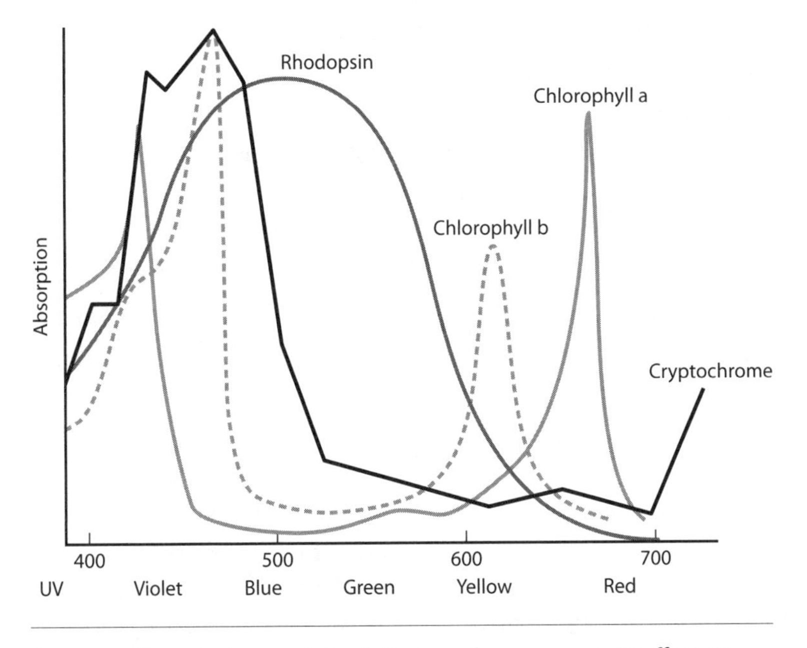

Spectrum of cryptochrome. Animals that use the paramagnetic effect created when photons generate paired radicals rely on a pigment in the retina. Only blue light is effective. Why this is true is made clear by the absorption spectrum of cryptochrome. For comparison the basic visual pigment of animals (rhodopsin) and the two main photosynthetic pigments of plants (chlorophyll *a* and *b*) are shown.

the air often contains bits of volcanic iron. In short, it's easy to find or generate magnetic material associated with animal samples, but showing that it is biological in origin is another matter. A few tricks familiar to geologists can help. For instance, the magnetic fields (*remanence*) arising from magnetite disappear at a very specific temperature (the Curie point). Heating readily distinguishes magnetite from the many inorganic sources of error. Another technique measures the field needed to reverse the polarity of the domains; biogenic magnetite tends to have a consistent grain size and thus quite a sharp reversal peak.

Honey bees have enormous numbers of superparamagnetic domains. According to the best-known histological study so far, the magnetite is confined to a specific set of innervated compartments in the upper part of the abdomen—tropocyte cells whose function had formerly been a mystery. There are about 8500 superparamagnetic grains in each tropocyte, embedded in a matrix of connective tissue. There are hundreds and hundreds of tropocytes per insect. The most obvious way to detect magnetic field strength and direction would be to measure the net force on the matrix. The structure will tend to swell across the axis of the earth's field as the parallel domains repel one another, while it will contract along the axis of the field as the head-to-tail interactions pull the grains toward each other. The strength of the expansion and contraction is a measure of the strength of the field. We calculate that the average bee has perhaps a hundred times as many superparamagnetic domains of magnetite as a perfectly designed sensory system would need to monitor the daily changes in the earth's magnetic field strength—the most exacting behavior known among invertebrates. As with the paramagnetic approach, superparamagnetic detectors cannot distinguish north from south in the horizontal plane; they must depend on detecting the dip direction. But unlike the cryptochrome system, these organs can work at night or in the complete darkness of the hive.

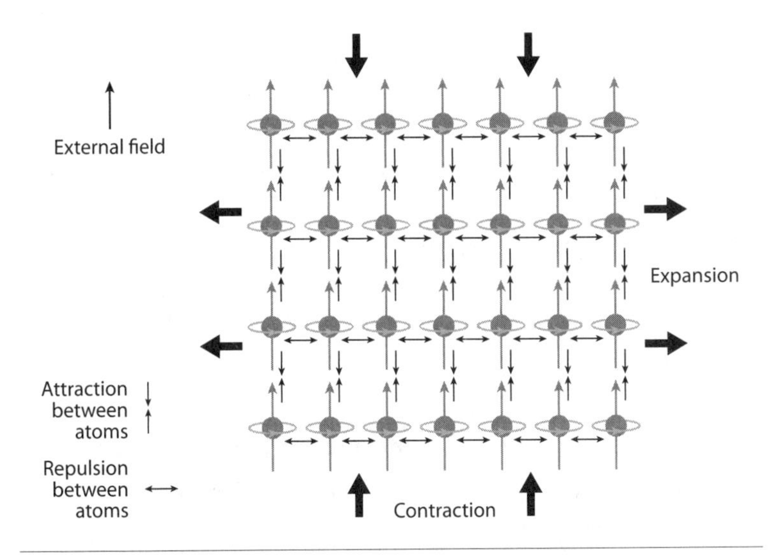

Forces in a superparamagnetic crystal. The spins of the unpaired electrons in a crystal too small to have a permanent self-stable magnetic field orient themselves to be parallel with the external (earth's) field. Atoms with spins aligned end to end attract each other, causing the crystal to contract along the north–south axis. Atoms to each side, on the other hand, repel one another, leading to expansion in the other two axes. Note that the response is the same for an external field pointed down: there is no ability to distinguish north from south; only the axis of the field can be measured.

Finally there is ferromagnetism, the permanent fields of large magnetite domains that we know as magnets. The first creature discovered to use magnets to orient was not an animal at all, but a bacterium. Microbiologist Richard Blakemore observed that the individuals from a mud-dwelling species swam toward the windows in his lab, apparently an unprecedented case of phototaxis in bacteria. More remarkable still, the same microorganisms swam *away* from the windows on the other side of the building. Blakemore realized that in both cases the bacteria were moving north and down; bringing a magnet next to them confirmed that these prokaryotes were traveling north along the earth's field lines. This species has a chain of 10–15 single-domain magnetite grains arranged in a line. The end-to-end arrangement of the magnetite creates a single magnet aligned with the long axis of the bacterium. The orientation is entirely passive: dead bacteria rotate into alignment with the external field as readily as live ones. The logic of the system is that the bacteria thrive in low-oxygen environments. Following the earth's field lines down (and incidentally to the north) returns the creatures to the oxygen-poor mud they seek.

Similar species in the Southern Hemisphere have magnetite chains with the opposite polarity and thus swim down and to the south. The evolution of this system probably began with a magnetite grain being used as a weight. As the densest substance cells can precipitate, it is an obvious candidate. Two grains would be even better. Once the number increases beyond two the end-to-end field will predominate. If these chains were formed randomly, half of the progeny bacteria would be swimming up to their deaths. But in fact the single long chain of the parent is split between the two daughter cells, preserving the polarity as a kind of cultural legacy.

Behaviorally, permanent magnets are distinguished by their ability to operate in the dark and their clear capacity to distinguish north from south. A long list of invertebrates have been shown to have magnetite, including monarch butterflies, spiny lobsters, migrating moths, and termites, which can orient underground. The permanent magnets of honey bees (of which they have many thousands) are located in the lower part of the abdomen along the midline. Like their superparamagnetic counterparts they begin forming in the larvae, and probably reach self stability during the pupal stage, when metamorphosis occurs. Pupation takes place in cells of the honeycomb such that each pupa is aligned perpendicularly to the axis of the comb. Presumably the permanent magnets that form then "fossilize" the field direction in the colony, facilitating the comb orientation behavior that follows swarming. Although there is no reason to assume that birds and bees evolved magnetic organs with the same division-of-labor logic, if convergent evolution is at work we might expect a high-sensitivity field-strength detector system separate from the low-sensitivity direction detector (compass). In birds, as we will see, the high-sensitivity system seems to be based on magnetite, whereas the compass depends on cryptochrome. In bees the superparamagnetic system is probably the high sensitivity strategy while the ferromagnetic approach might deal with compass tasks.

As we have seen, invertebrates are equipped with backup systems to keep them well oriented in diverse situations. Diurnal insects focus on the sun, employing clever tricks for locating it in the sky despite being legally blind. They calibrate the sun's direction of travel, carefully estimating, extrapolating, or memorizing its rate of azimuth movement as appropriate. They use polarized light patterns when possible to locate the sun when it is hidden, incorporate landmarks to judge azimuth, and fill in navigational gaps as needed. Finally, they call upon the earth's magnetic field when all else fails. Vertebrates, with their high-resolution vision, larger home ranges, and (for some species) epic migrations augment these strategies with new compasses and more elaborate processing to survive the challenges of their more expansive world.

Chapter 5

Vertebrate Compasses

Though the small, orange-breasted European robin is a popular symbol of Christmas, many populations actually overwinter not in Europe, but in northern Africa. Robins are famous for their friendliness toward humans, but also for the males' readiness to fight other males—or anything else orange—in the spring. The robin is a classic instance of how internal timers orchestrate behavior. In the winter, as the days grow longer after the solstice, cells in the brain start secreting hormones such as prolactin. These hormones cause migratory robins to begin eating more and building up fat stores, well in advance of any need or promising change in the weather. A month later even the birds that don't migrate are seized by *zugunruhe*, or migratory restlessness, and become hyperactive at night. Other hormones, secreted when the days have lengthened yet further, cause the males to develop breeding plumage—the orange other males find so provoking.

Soon the birds in migratory populations take wing, flying hundreds of miles north to where they were hatched. On their breeding grounds the males become intensely territorial, attacking other males and courting females. Pairing, however, causes the male to switch his priorities to building a nest; the female lays an egg a day until she is satisfied with the total, and then the pair begins peacefully incubating. Once their brood has hatched, they work tirelessly to feed the offspring. When the shortening days of late summer trigger a new round of hormonal changes and reshuffling of priorities, the robins prepare to redeploy back to northern Africa. With internal timers choreographing migrational motivation, the first question for navigation researchers is how these long-distance travelers know which direction (or, more often, sequence of directions) to fly, and how they find that bearing despite everything the weather can do to obscure any potential cues.

Techniques

Much of what we know about invertebrate orientation came initially from honey bees. They are the lab rats of navigation, steering their way patiently to and from food sources hour after hour, sketching out scaled-down maps after each journey. They put up with being numbered, carrying weights and flaps, having their hive redesigned and reoriented for experimenters' convenience, and seeing crude artificial cues substituted inexplicably for the ones the natural world provides. Research on vertebrate navigation has followed the same course. A few tractable and convenient species matched with investigators' imagination and technical prowess can provide a window into the navigational computer of, in particular, sea turtles and birds.

Though long-distance migrators like the European robin provide data only twice a year en route to summer or winter quarters, they are irreplaceable sources of information. Robins, warblers, and sparrows in particular can be held at least briefly in cages, and their fluttering attempts to escape reveal the direction they long to fly. One way of measuring their directional tendencies is to put the birds singly into donut-shaped enclosures with an array of computerized radial perches. Most migration occurs at night, and in the evening the birds become highly active, hopping back and forth on the perches on the side of the cage closest to the direction that unrestrained migrants are flying overhead. Another way of measuring their directional preference is to place the birds in large paper cones that make flight difficult, and place an ink pad at the bottom. The birds try to flutter out, leaving footprints on the cone in the process. Later a scanner registers the density of the marks and determines the mean direction in which the bird is trying to escape. Either technique allows the researcher to artificially manipulate the local compass cues experienced by birds in the apparatus.

While cage tests can provide several nights of data from a limited number of birds, other approaches can expand the sample sizes, the number of locations, or the distances over which data are taken. Short-wavelength radar returns echoes from birds passing overhead, a fact discovered during the Second World War when migrants periodically obscured military targets. Radar data can tell us when and in what direction animals choose to migrate, how they deal with crosswinds and overcast, and a variety of other

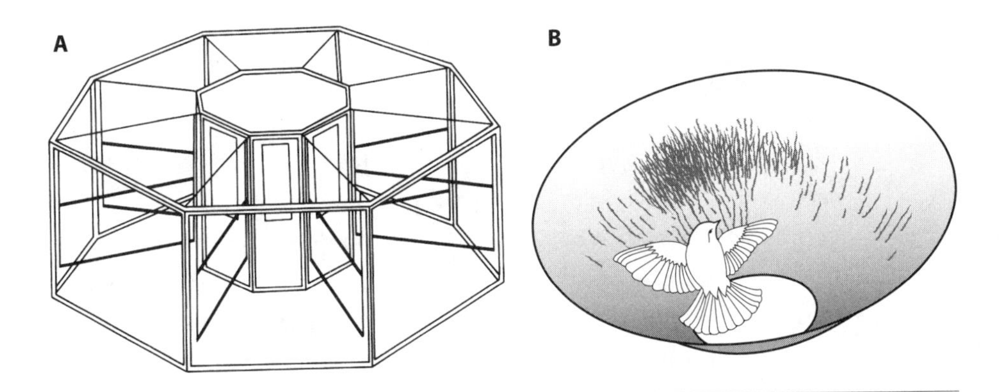

Recording nocturnal orientation in birds. (A) The Kramer cage is a donut-shaped enclosure with radial perches. Birds hop around the periphery, most often on the side facing the preferred migratory orientation. (B) Each time the would-be migrant attempts to escape an Emlen cone it leaves behind an inky set of footprints. The marks are densest in the preferred direction for migration observed in free-flying birds.

things that cage tests cannot address. Of course manipulating compass cues for birds thousands of feet above is out of the question, and the exact distribution of species can only be inferred.

More recently the size and power requirements of miniature radio transmitters have decreased to the point that some moderate-sized birds can carry fairly long-lived digital devices. Most of the weight is in the battery, which must remain light, so there is a trade-off between the duration of the signal, its range, the frequency of transmission, internal processing and monitoring needs, and other power-based practicalities. Tracking devices for large birds such as hawks or certain seabirds include radios powerful enough to signal orbiting satellites in real time over a matter of months. With medium-sized animals the circuitry more often records GPS position at frequent intervals and stores the data for later short-range broadcast. Smaller birds can carry only lightweight recorders that store data or merely log the times of sunset and sunrise, allowing researchers to compute latitude and longitude—if they can manage to recover the devices.

■ Homing Pigeons

Whatever the technique, migrants are limited to two performances each year. Clearly researchers would prefer a species that will undertake long-distance navigational tasks on demand. The closest thing to a honey bee in the bird world is the homing pigeon, a domesticated version of the rock dove. Experiments with these hardy and reliable subjects provide information about the analogous navigational systems of migrating species. Researchers use homing pigeons to discover what compass senses are available, what a bird does when its primary orientational cue vanishes, and how it compensates for the passage of time. Pigeons can tell us how much birds know about their location, and how the use of any internal compass might change with age and experience. Detailed information from pigeons has allowed researchers to pose the sorts of focused questions that are essential but difficult to ask when working with wide-ranging, twice-a-year migrants.

The common rock dove is an incredibly successful species, having spread throughout the world from an already extensive prehuman range stretching from the Mediterranean to India. These birds are opportunistic breeders, ready to reproduce year-round whenever conditions are favorable. This trait makes them perfect for breeding in labs for psychological research, or raising in dovecotes for food or in lofts by pigeon fanciers keeping birds simply for racing against the homers in competing lofts. The monogamous pairs can fledge a pair of squabs about every seven weeks, feeding them with a unique secretion called crop milk; this mammal-like strategy is seen in no other group of birds. Rock doves are moderately social, greatly preferring colonial nesting and foraging in flocks—habits bred into them by selection as a way of having many eyes at once looking out for their archenemies, peregrine falcons and sparrowhawks. For a nesting site they need nothing more than a small shelf on a cliff, a windowsill, or an ashtray in a cage.

Pigeons are sociable and domestic birds, flying out to forage for seeds each day and returning home for the night. Their love for their home loft has been noted and put to use for millennia; the earliest records of their homing abilities date back 3000 years to Egypt and Persia (now Iran). Centuries of breeding by pigeon fanciers, Darwin among them, have resulted in birds that will return home quickly and directly after being transported up to 600 miles. They are well oriented toward their loft even at much greater distances, but few domestic pigeons can manage the physiological challenge of traveling so far. Though the very breeding that has hypertrophied their navigational prowess may make them less than ideal models for wild migrants, most researchers in the field see no fundamental difference between the map and compass abilities of pigeons and those of conventional migrants. Allowing for the specializations that distinguish the thousands of kinds of birds, each adapted to its own niche and associated set of challenges, the navigational system of pigeons seems pretty generic.

Pigeon racers allow their birds to fly free near their home loft, and then train them through releases from increasing distances. At a racing meet, hundreds of birds are transported to the release site and allowed to take wing simultaneously. Their arrival times at their home lofts are recorded and their velocities calculated. In a typical scientific release, on the other hand, perhaps 30 pigeons are carried in cages or baskets, usually with no view of their surroundings, to a release site new to them. Some of the birds will be subjected to an experimental treatment either before, during, or after transport. Researchers toss one bird at a time, alternating an experimental with a control bird at roughly 10-minute intervals. Most pigeons circle the release site two or three times as though they are getting their bearings. Given a choice of cage doors to exit, pigeons favor the one opening toward home, indicating that even flight is unnecessary for some degree of initial orientation. The direction in which each bird disappears from sight is its *vanishing bearing*.

Once one bird disappears, the next can be tossed. Release sites are carefully chosen to be high, open points with good visibility, so

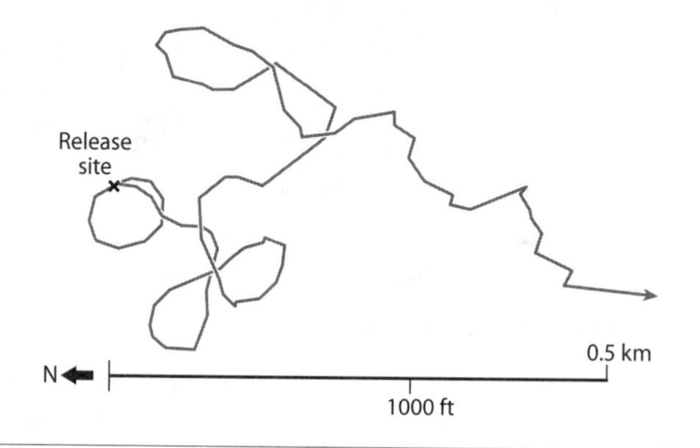

Track of bird just after release. As is typical for pigeons tossed at a release site, the animal completes a few loops before setting off for home. Note that from the outset the bird is drifting in the correct direction. A highly motivated bird may dispense with this initial circling and head straight for the loft.

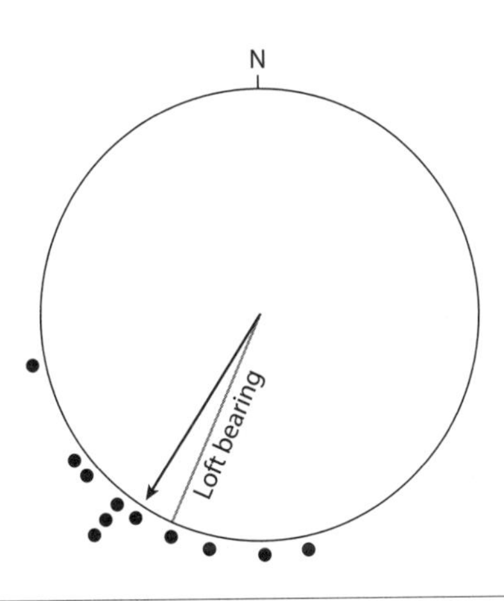

A vanishing bearing record. Each point on the periphery represents the vanishing bearing of a single bird. The loft direction is shown as a line. The arrow is the mean vector, indicating by its angle the mean direction of the bearings and by its length the degree of clustering. If all bearings were at the same angle the arrow would touch the circle and have a length of 1.0; if they were randomly scattered the arrow would have a length of 0.0.

the pigeons can be observed for a long way before being lost to sight behind trees or hills. The longer a pigeon flies away from the site before disappearing, the more tightly clustered are the vanishing bearings. In part this improvement is an illusion created by parallax.

Radio transmitters on the birds allow researchers to track the vanishing bearings farther. What these bearings show is that the birds, though they show a strong tendency to fly toward home from the start, become more certain of their location as they travel, and become more precisely directional. It's important not to be carried away with initial pigeon accuracy: even a relatively tightly clustered data set is typically spread 15° left and right of the homeward direction. For a release even at just 50 miles from the home loft, this corresponds to a range of apparent initial uncertainty about the location of release on the order of 25 miles. Radio tracking makes it possible to monitor returning birds as they near the

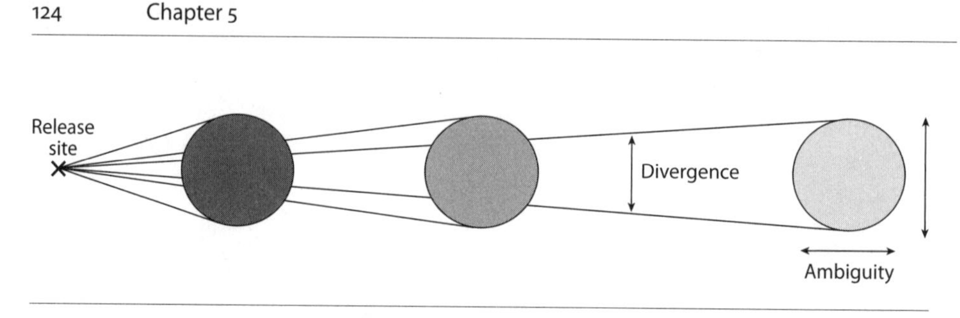

Effect of parallax on apparent precision. A given error (or *ambiguity*) yields an ever-smaller angular deviation with increasing distance. Thus, vanishing bearings automatically appear more precise the farther away the birds travel.

loft. In a classic test, Klaus Schmidt-Koenig of the University of Tübingen equipped pigeons with frosted goggles that deprived them of form vision. Triangulation of their radio bearings showed that the returning birds got surprisingly close to home, within perhaps a mile. This remains the single best estimate of the accuracy of pigeon position sense after at least 30 minutes in the air.

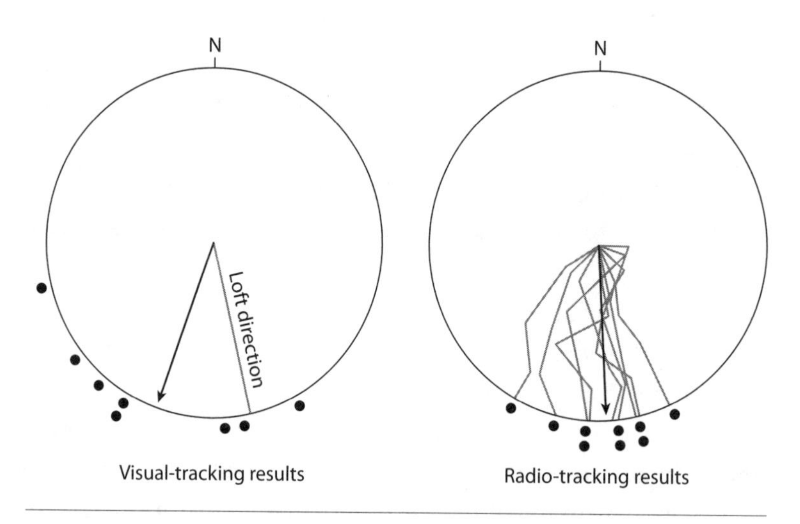

Visual and radio vanishing bearings compared. Radio-tracked birds disappeared at about five times the distance from the release site as visually tracked pigeons. They are more clustered (yielding a longer mean vector) in part as a result of the parallax effect. That the bearing is closer to that of the loft direction, however, suggests that their compass orientation or sense of location (or both) also has improved with distance.

The heroic task of following pigeons en route home used to rely on radio tracking from a plane shadowing each bird. The development of GPS data-logging technology has permitted higher-resolution mapping of the paths taken back home. This technology has revealed such unexpected insights as the marked effects man-made landmarks can have on the birds and the interactions birds sometimes have with other birds on the wing. The main lessons are that the birds' routes are not straight lines, do not necessarily point directly toward home, and seem to be subject to midcourse correc-

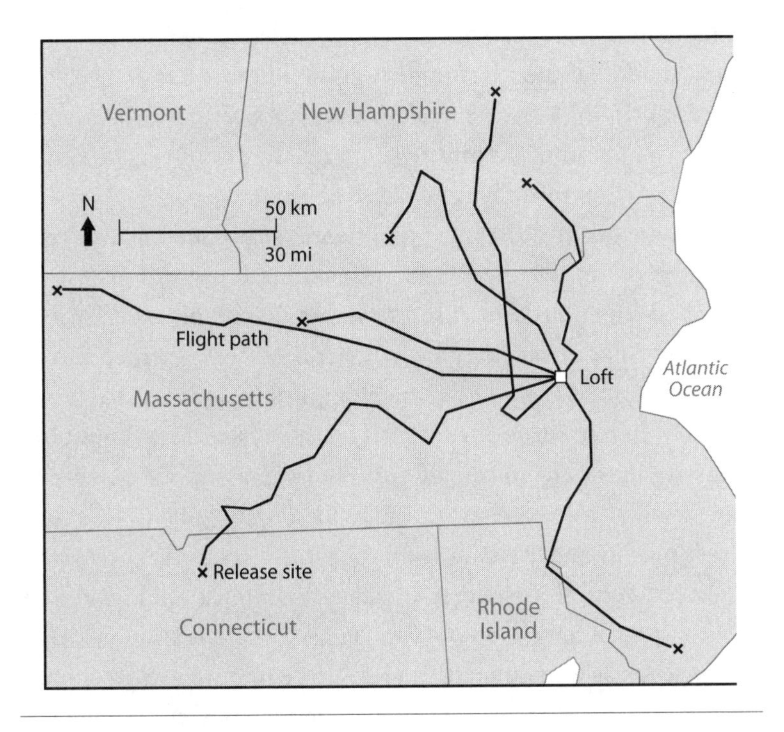

Homing tracks of pigeons. Birds from the same loft were released and tracked by air from a variety of novel sites 50–100 miles from the loft. One bird clearly set off in the wrong direction, whereas others displayed relatively small initial errors. Most appeared to make at least minor midcourse corrections, implying they were updating their navigational information en route. One made a major correction about 15 miles from home, well outside visual range of the loft.

tions. The chapter about true maps will discuss what each of these trends tells us.

In the wild, young pigeons fledge at about five weeks and begin taking short excursions around the nest that range out to perhaps half a mile. These practice flights help in at least three ways: they build endurance, they lead to an ability to recognize home from nearby, and they allow the birds to learn something general about their home's location in the larger world. It is at this point that fledglings imprint on, or memorize, their home nest or loft. Taken to a new home after just one or two brief outings and confined for up to several years, pigeons will nevertheless generally return to their natal loft at the first opportunity. Without these early excursions, imprinting either is weak or does not occur. Most researchers attempt to mimic natural development by permitting free entry to and exit from the loft, at least during the day.

A dramatically different experimental technique involves rearing the birds inside the loft from hatching onward. For testing, these first-flight birds are taken from home with no experience in free flight whatsoever. They are in general less well oriented and far less likely to find their way home than birds allowed to come and go freely. In one study, for instance, only 0.5% of first-flight birds released 55 miles from the loft returned successfully; by contrast, after being allowed a few local practice flights fully 50% of first-release pigeons make it back from the same distance. (Success rates for second- and third-release practice-flight birds are higher still, rapidly approaching 100%.) As we will see, there is good evidence that pigeons as well as many long-distance migrants learn enough during early flights to make major improvements in their navigational strategy. Whether homing pigeons or conventional migrants, birds not permitted the opportunity to gain relevant experience cannot mature in the normal way. Pigeons subject to this sort of unnatural close confinement through their impressionable adolescence may be developing an alternative set of orientational

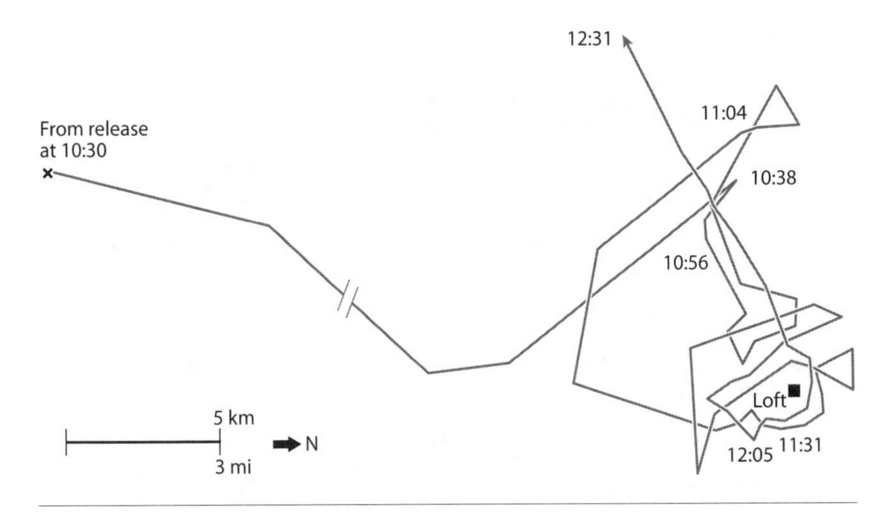

Tracks of a visually impaired bird near its loft. The pigeon is wearing frosted goggles, which eliminate form vision without preventing the animal from at least roughly judging the sun's direction. Despite this handicap the bird circles within a quarter mile of the loft. This establishes the precision of its navigation, and defines the radius within which it needs to be able to recognize the loft and its surroundings visually in order to land in the right place.

priorities. This rearing technique seems to encourage the birds to focus on airborne odors, as we will see when we look at true maps.

The pigeons so intensively studied by orientation researchers are the descendants of generation after generation of highly selected racing birds that may or may not have improved orientation skills as a result of their breeding. What is clear is that most ordinary species of birds are as a rule quite poorly equipped for displacement studies, and do not get back very often after even modest relocation. But this may just be a matter of motivation: during migration season they too can be transported hundreds or even thousands of miles and then either find their way home or reorient to and continue on to their intended goal with a high degree of success.

■ Beacons

Animals with a classical compass are sensing an absolute direction such as magnetic north and then using this standard to adopt the proper orientation—ESE, for instance. The desired direction may be learned or innate or calculated; it may or may not require time compensation. The key thing is that a larger coordinate system is involved, and reading the compass is only the first of at least two steps. But vertebrates sometimes avoid putting the navigational problem into this larger context. Instead, they make use of *beacons* in ways that are commonplace and often quite sophisticated. For example, sea turtles hatching from their buried eggs at night on a gently sloping beach must get away from terrestrial predators and into the water as quickly as possible. But where is the shoreline? Hatchlings scan the horizon and set off in the least-dark direction—almost always the horizon over the ocean. Place an artificial light on the beach and they will head for it instead.

Once a hatchling sets out across the beach, it can maintain its direction because the beacon remains in place; there is no danger it will find itself walking in circles in the dark. But as soon as the baby turtle reaches the ocean, the situation begins to change. Its head barely clears the water, and within a few minutes it is surrounded by horizon. How is it to keep from swimming an arc and arriving back on the beach? The moment the turtle enters the water, it switches cues: instead of heading toward light it turns head-on into the waves. Wave direction becomes the beacon that leads the animal away from land and into the sea. Once the escape is well under way, the turtles begin using more conventional compasses to guide their journey. In that first hour the hatchlings take magnetic bearings of both light and wave direction and, making the transition from beacon to compass, steer this course until far from shore.

Perhaps the most remarkable use of beacons is seen in salmon, the dozen or so members of the trout family that migrate from inland spawning sites to the ocean and then back to the same stream years later. Depending on the species, hatchling salmon spend a few months to three years in freshwater before maturing into juveniles, or *smolts*, and heading downstream. After a pause in brackish water near the coast where they acclimate to the very different body chemistry required for the salt water of the ocean, they swim out to sea for one to five years. Their return to their natal streams involves a map sense based on information collected when they first entered the ocean, which in turn calls into play both celestial and magnetic compass information to orient the marine phase of their trip back. Once in the river they home in on an olfactory beacon, the odors memorized on their first day as smolts. The salmon follow this beacon doggedly; some breeding in Idaho, for instance, travel 900 miles and climb 7000 feet to locate the tiny creek they hatched in years before.

This rigidly programmed use of olfactory beacons is unusual; beacons are more typically employed as a stopgap to maintain a consistent bearing when a compass cue is temporarily absent. The challenge is to continue moving in a straight line in the intended direction. This means compensating for drift from both crosswinds and currents, and defeating the inherent tendency of all animals toward flying, walking, or swimming in circles.

A bird with a clear view of the ground during the day, a good estimate of altitude, and the appropriate neural machinery should be able to compare the direction of apparent movement of the ground with its actual heading and air speed, and from this compute its drift. With this parameter in hand the bird can compensate for the wind's direction and speed. At sea, where the visibility down and around is often quite limited, animals face a more difficult prospect.

It was once thought that animals might keep precise track of their heading and route through inertial measurements, thereby avoiding both drift and circling. Like the advanced autopilots used in airplanes before GPS satellites made them redundant, the system would need to keep track of every acceleration event—that is, every change in speed or direction. Airplanes had gyroscopes and accelerometers. Avian and mammalian inner ears have instead a vestibular system, which includes three semicircular canals at right angles to each other, plus three otolith organs. *Otoliths* are small, bonelike particles exquisitely sensitive to pressure and angle. Any change in movement—that is to say, any acceleration—induces fluid flow in the canals. The otoliths respond primarily to gravity, giving the canals a reference point. Because speed is the integral of acceleration, and distance is the integral of speed, a sufficiently precise set of three-axis accelerometers could in theory allow a creature to compute not only present heading, but also the net effect of changing crosswinds. (An animal does not feel a crosswind in real time but only the lateral acceleration when it was first encountered.)

As we will see when we look at local navigation and mapping, there is good reason to think humans and other animals can keep track of displacements inertially for a few minutes or even hours if conditions are right. Inexperienced pigeons less than 10–12 weeks old depend in large part on inertial and other information gathered during displacement. But in the long run, organisms forced to depend on inertial guidance and deprived of compass information inevitably trace large circles. The size and direction of the circle (CW vs. CCW) depend on the individual asymmetries of the animal. Even with a compass to maintain a consistent bearing, the animals slowly lose track of position on their outward journey and drift farther and farther off course when they attempt to return. Inertial navigation is simply not a reliable antidote for circling biases or crosswinds and currents on longer trips. Moreover, it seems that experienced pigeons ignore inertial information: transport under complete anesthesia or after surgical severing of the most important of the semicircular canals has no effect whatsoever on their homing ability.

That day-flying insects can compensate for drift has been clear since the first discovery of the ability in honey bees. That the high-altitude, nocturnal-migrating moths in the previous chapter can do the same is slightly puzzling. The moths take to the air if the breeze is in approximately the right direction and then adjust for the deviation in this tailwind to fly a true migratory course. But once they are airborne at night, how can they detect the discrepancy between their heading and their ground track? They need to see the earth. Either their low-light vision is exceptional, requiring only starlight, or they are using ground lights as beacons. By focusing on a light dead ahead and tracking its apparent drift to the right or left, the moth can adjust its heading to correct for the crosswind component in the tailwind. Recent theoretical calculations suggest that moths and birds ought to be able to detect patterns of turbulence in the air through which they are flying, and use this information to infer wind direction. As of now, however, this remains only a promising idea.

For birds that fly at moderate elevations, there is good evidence that while a direct view of the ground or terrestrial lights is unnecessary, conventional cues are sufficient to account for wind compensation. In a heroic set of experiments, Donald Griffin, the brilliant animal behaviorist who discovered bat echolocation and popularized the study of animal intelligence, tracked migrating birds on cloudy nights with a war-surplus antiaircraft radar and a small team of shivering students. We spent many a cold night helping with the migration project. In addition to using measurements of conventional cloud height and thickness provided each hour by surrounding airports, we flew a cloud detector on a small blimp, measuring cloud opacity over a vertical transect extending for hundreds of feet above and below the altitude at which the birds were flying, thus making sure that the birds could not see the stars or the ground. Griffin found that, like moths, birds wait for favorable winds and fly a true course through overcast and between clouds.

If the birds in Griffin's study were not using ground lights as beacons, what was their strategy? Griffin used the same blimp to raise a variety of instruments up to check. The answer for migrants traveling at modest heights in the spring may involve auditory beacons, such as isolated ponds with calling frogs. By concentrating on a cluster of chorusing frogs straight ahead, a bird can sense crosswind drift and make the necessary adjustments.

For the many migrants flying higher, something louder and more distant is probably needed. If it's an acoustic beacon again, the best bet seems to be infrasound. Lower frequencies travel farther than higher-pitched ones, whether through the water or the air. It is the selective loss of high frequencies that gives familiar sounds a faraway quality, and allows us to judge acoustic distance. Homing pigeons (and presumably other birds) can hear what are for us subsonic sounds—rumblings below 20 Hz, which we can only sense as vibrations from objects we are touching. In our history as a species there has been no selection for hearing below 20 Hz or above 20,000 Hz, and thus the elaborate specializations necessary for this ability have not evolved. For creatures such as bats, mice, moths, and owls on the other hand, the ultrasonic region above our range of hearing is a familiar and essential part of daily experience. For elephants and blue whales, subsonic sounds communicate the whereabouts of others of their kind.

You might wonder how anyone knows about infrasonic sensitivity in birds. As with many putative senses, discovery depends on finding the right behavior to monitor. In many cases responses to stimuli depend critically on the context. Pigeons, for instance, can learn to associate colors with food, but not sounds. If the task is learning about danger, they remember sounds but not colors. Rats and a variety of other creatures have their own biases. Given that pigeons live on seeds, remembering the sounds associated with a food item would be useless at best. In short, animals often come to learning situations with strong context-specific biases. When we ask a caged bird in the laboratory whether it can sense low-frequency sounds or changes in atmospheric pressure or odors in the air, we have little hope of getting a positive answer unless we tap into a relevant behavior. If that response involves actually being airborne, the difficulty of the experiment increases exponentially.

Fortunately for researchers, there are involuntary reactions that are context-free. An unexpected movement, flash of light, sharp noise, or other change will often grab our attention. In tests to determine whether babies recognize consonant sounds, for instance, a highly reliable technique is to play one sound over and over until the child's attention wanders, which happens very quickly, and then make a small change to the stimulus. If the infant detects the change, it will look immediately at the source. This reaction is even clearer if heart rate is monitored instead, because there is almost always a momentary acceleration or deceleration after a perceptible change. Independent of what a stimulus means to them, birds will show their sensitivity to a cue by involuntarily adding or skipping a heartbeat.

Given that they can hear infrasound, what have pigeons evolved to listen to in this deep bass register? No one really knows for sure, but meteorologists use infrasonic detectors to localize and track distant thunderstorms (admittedly at frequencies below even the range of bird hearing). More relevantly, a variety of infrasonic beacons have been detected, such as the low-pitched whistling of wind passing over mountain ranges. When birds are flying through clouds, such acoustic landmarks might provide essential reference points for correcting what could otherwise be a fatal drift. Localizing low-frequency sound is difficult, as many of us know from experience. For a fast-flying bird listening to a familiar sound, its Doppler shift might provide the simplest clue.

■ The Sun and Sky

Honey bees, as we've seen, know how the sun moves through the sky. Some of the time this ability is based on a memory of the sun's path on previous days, but foragers also can infer the track for parts of the day during which they've never seen the sun. Can we expect anything less of vertebrates?

The classic demonstrations of a true time-compensated avian sun compass involve clock shifts in which birds are kept indoors for a time while the artificial dawn and dusk are shifted out of phase with the outside world—generally six hours fast or slow. Birds in these tests do not view the sun during the clock-shifting process; lights-on and lights-off in the lab are the only cues available. Researchers record the direction that migrating birds are hopping, or the vanishing bearings of homing pigeons.

It's worth considering what the birds see in the sky after such a manipulation. Assume that the actual time is noon, and that home is in the south. A properly oriented pigeon will fly south toward the sun (or the corresponding pattern of polarized light centered on the sun) high in the midday sky. If the release had been at 6 a.m., the same unshifted control bird (assuming it understood the sun's arc in the sky for the appropriate date and latitude) would judge the sun to be in the ENE and fly about 105° to the right of the rising sun, still low in the sky. Now consider a bird shifted six hours early, whose internal clock reads 6 a.m. The sun is in fact in the south at an impossibly high elevation for that hour, but nevertheless the animal takes its azimuth to be more or less in the east and sets off well to the right of it, flying to the west. Similarly, a bird shifted six hours late interprets the noontime sun to be setting in the WNW and strikes out well to the left of it, heading in fact east.

Initially clock-shift tests with birds asked only qualitatively whether departure directions were altered by the treatment and if so, whether the reorientation was in the expected direction. The answer is generally yes, though odd things sometimes happened if the shift treatment went on for too long prior to the test. An anthropocentric assumption that animal systems must be simple underlay much orientation work at the time, and animals were not expected to understand too much nor be inclined to make complex navigational calculations. They might know that the sun

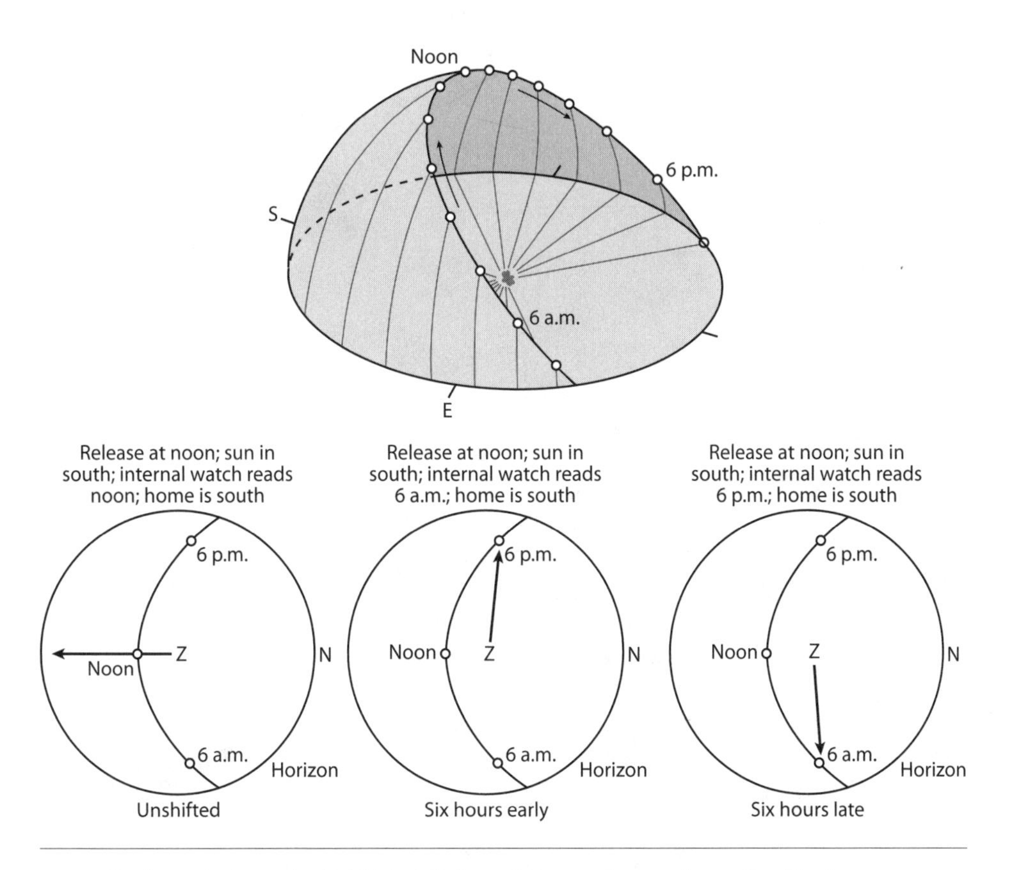

Sun's arc during a clock-shift test. Released at noon from a site north of their loft, pigeons that are able to compensate for the sun's westward motion through the sky should fly in one of three directions: unshifted controls should set off to the south; birds shifted six hours early should adopt a westward heading; and pigeons shifted six hours late should depart more or less to the east. Note that north is to the right in each figure; Z = zenith.

moves, but could hardly be expected to grasp the details of spherical geometry. After all, they seemed oblivious to the enormous incompatibility between the expected and actual elevation of the sun after a shift.

More recently, in light of the discoveries with insects, researchers have begun to look at clock-shift results more quantitatively. If birds, like their invertebrate counterparts, actually memorize or

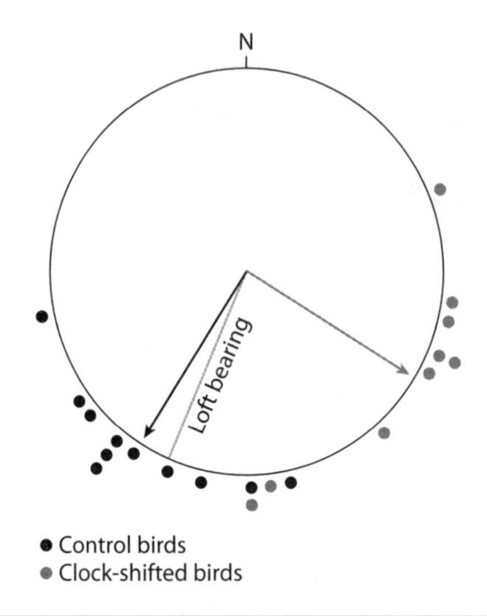

Results of a clock-shift test. The loft is SSW of the release site. Unshifted (control) birds are well oriented, whereas birds shifted six hours late depart roughly 90° to the left of home. Note that the experimental birds are less clustered (as indicated by the shorter mean vector), which is largely a consequence of two birds that flew south, roughly in the correct direction. This seeming dichotomy is not unusual: these may be younger birds that have not yet learned to use the sun accurately. Despite the shift, all of the pigeons reached home.

compute the sun's shifting arc on a daily basis, then they ought to be able to compensate differentially for the relatively slow change in azimuth near dawn and dusk, as well as the unusually rapid shift near noon. After shorter (two-, three-, or four-hour) shifts, pigeons do indeed demonstrate their ability to adjust to the sun's daily azimuth acceleration and subsequent slowing.

But though the sun (or the polarized light pattern it generates) seems the obvious compass cue for pigeons to use, research into the ontogeny of this orientation system suggests a different initial hierarchy. Pigeons fledge at about 5 weeks of age, yet do not display precise sun-compass use until roughly 10–12 weeks. During this period their sensitivity to clock shifting gradually increases from

0% to 100%, another indication of a tactical shift. But despite this major reconfiguration of their orientation processing, pigeons are able to home with modest success even when just a month and a half old (though they improve greatly with age and experience). How can this be possible?

We have to keep in mind that there are two things going on with pigeons and many migrants. In true navigation as we defined it in the first chapter, the animal knows where it is relative to home (and in some cases, it would seem, its absolute position on the planet); from that knowledge it computes a return bearing, which it must then find with a compass. The change in pigeon strategies involves both the map step and the compass. The youngest homers use the inertial mapping strategy (keeping track of the journey's outward legs in the manner of dead reckoning) plus other cues experienced en route to determine location, then use a compass that has no provision for time compensation to set their course home. At about 10 weeks of age—sooner with experience flying farther from the loft—they begin to shift automatically to a true map sense (the ability to figure out where they are from information at the release site itself) and, on clear days, use a time-compensated celestial compass to aim their return flight.

Migrating birds also seem to change strategies with age or experience, especially between their trip south that first autumn and their later journeys, both north and south. One population of song sparrows, for instance, flies a thousand miles south from their breeding grounds in British Columbia to winter around the Gulf of California. If first-year birds are transported 2000 miles east across the continent and released, they fly directly south, rather than the SW course that would take them to their wintering grounds in Mexico. Second-year and older birds, however, fly SW. While the purpose of these strategic changes would seem to be to make things easier for the youngest animals (who are most at risk of getting lost), the details are sufficiently complex to have kept researchers in some confusion.

Any system that the birds might use as a time-independent, inborn compass must provide relatively unambiguous directional information with no need for detailed learning or calibration. One possibility is the pattern of polarized light at sunrise and sunset. At these specific moments the band of maximum polarization passes directly overhead. At the zenith, the angle of polarization is exactly perpendicular to the sun's azimuth. By observing this angle at dawn and dusk a bird can determine true north and south, the directions midway between the rising and setting of the sun. Because most migrants take wing near dusk and fly at night, it's perhaps not surprising that they are particularly attentive to the sky at sunset. Rotating the pattern of polarized light overhead redirects their initial orientation, showing that they are ignoring the sun itself (or the bright spot it leaves on the horizon).

Another absolute in the celestial hemisphere is the *pole point*, the exact spot in the sky about which the sun, moon, stars, and polarized light patterns appear to rotate. This is easiest to visualize by looking at a time-lapse picture of the night sky. The stars trace out arcs centered on the pole point, which is located due north at an elevation that corresponds to the latitude—60° for Oslo, 51.5° for London, 40.7° for New York, 29.7° for Houston, and so on. Taking a bearing on the North Star was for centuries the most precise way to judge direction and latitude, but Polaris is not exactly at the pole point. It is located about 0.7° off axis, and the precise value is changing. In 13,000 years—an instant in evolutionary time—Polaris will be fully 23.4° away from the center of celestial rotation. Over much of that interval there will be no bright star near this critical spot in the sky. (Polaris has no Southern Hemisphere equivalent.)

Though Polaris itself may be hidden by a cloud, the general patterns—the constellations the stars appear to form—glimpsed through gaps in the clouds can be used to infer the position of the pole point. Indeed, most of us generally locate the North Star by

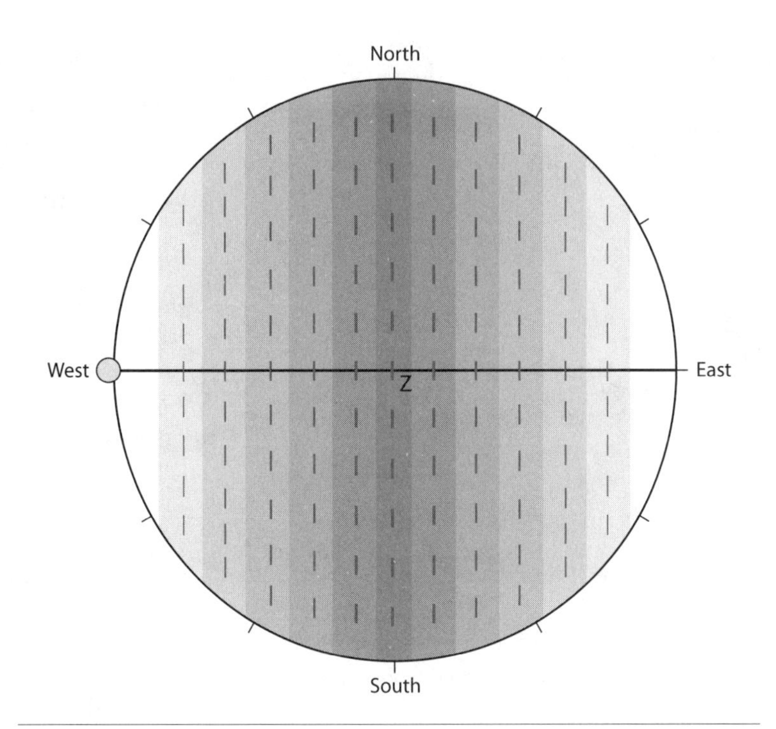

Polarization at dusk. Because the band of maximum polarization is strongest 90° from the sun, sunrise and sunset are unique times. At these two moments the band of maximum polarization passes through the zenith directly overhead. Because the angle of polarization is always perpendicular to the sun's direction, the zenith point also reveals the sun's location as it slips below the horizon. The figure shows sunset on the equinox, on which date the sun is due west.

sighting along a pair of stars that form the outer edge of the imaginary cup in the Big Dipper—or the cutting edge of the Plough, or the middle rib in Ursa Major (The Great Bear), depending on what your eye sees and your brain fabricates in the sky. If you've learned the constellations, and especially if you can see two that are widely separated, inferring or even triangulating the pole point is fairly easy.

Star tracks. The movement of the stars over the several hours of this time-lapse photograph circles around the pole point. The star closest to the center is Polaris, about 0.7° away from the center of celestial rotation at the moment. This photograph was taken from the Canary Islands, at roughly 30° latitude; thus the pole point is 30° above the horizon.

The same process seems to be exploited by many nocturnal migrants. Shortly after fledging juvenile birds memorize the star patterns overhead, paying particular attention to the pole. Elegant planetarium experiments by Steve Emlen at Cornell show that the birds identify the pole point by the pattern of rotation; it can be anywhere in the sky. In addition, any configuration of stars will do; the birds memorize whatever artificial pattern they see, again focusing on the pole point. But in contrast to the effect of conventional imprinting, which is irreversible, these animals can update their celestial snapshot. This is essential because as the earth circles the sun, many of the spring constellations disappear—that is, they begin to rise earlier and earlier (four minutes each day) until they are above the horizon only during daylight. A different set of (autumn) constellations slowly becomes visible. Birds add stars to their sky maps throughout the year. (There is no evidence that migrants see the sky as a collection of constellations; this intellectual shortcut may be uniquely human.)

Yet another difficulty confronts young migrants. As they move south in their first year, some northerly constellations disappear below the horizon as the elevation of the pole point declines with latitude, and new ones appear in the south. Some extreme long-distance migrants move from the northern Arctic to near the southern polar zone, and in the process must commit to memory most of the brighter stars visible from our planet.

The pole point is visible in the daylight sky as well—if, that is, you can see polarized light and detect its motion. The polarization at this unique spot in the sky rotates more tightly here than anywhere else. To an animal sensitive to this kind of movement it would appear to be something like a flashing light in the heavens.

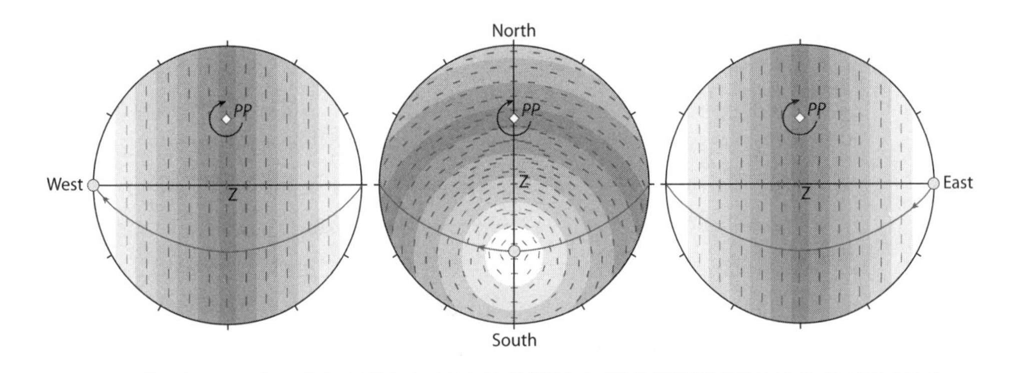

The pole point in the daytime sky. The sky as seen on the equinox at 45° latitude in the Northern Hemisphere. The sun rises due east (right), traces the arc shown, reaches its highest elevation at solar noon (center), and sets due west (left). Throughout the day the pattern of polarization circles the pole point (*PP*); this is also the location in the sky with the maximum rate of rotation.

■ Magnetic Fields

As we've seen, migrating birds are pretty well equipped to travel when at least part of the daytime or night sky is visible. But much of the time (and particularly during spring migration in much of North America and Europe) the sky is overcast. Fronts build up and move rapidly, rainy days are numerous, and temperatures can change abruptly. Robins, warblers, and sparrows have dutifully abandoned their tropical winter quarters to reach their nesting grounds and be the first to grab the best territories. What are they to do without the sun and stars en route? Until radar, many researchers assumed they bided their time at rest stops along the way, waiting for the weather to break. But though traffic is reduced in cloudy weather, migrants are still on the move. Our tedious measurements of cloud density, moving a small detector up and down on a blimp, convinced us that celestial cues were dispensable. Longer-range radars (such as the satellite tracker at Wallop's Island that NASA was kind enough to loan) confirmed that the same pattern holds at higher altitudes and over a much wider area. Calling frogs can hardly account for the birds' ability to navigate without their celestial compass.

In fact, the intellectual problem is more serious now that we know that pigeons orient accurately long before they learn the sun's arc. To do this they must rely for a few weeks on a time-independent reference compass or cue. But there were strong hints of this much earlier. By the 1950s, for instance, clock-shift experiments indicated that jet-lagged birds are not fooled under overcast. This remarkable and deeply puzzling result proved that migrants and pigeons substitute a backup compass when the sun is not available. What could this occult sense be?

Researchers considered many exotic suggestions for this secondary compass. One popular long shot was *Coriolis force*—the effect created by the different speed of the earth's rotation at different latitudes. This force is both invisible and (at the scale of an ani-

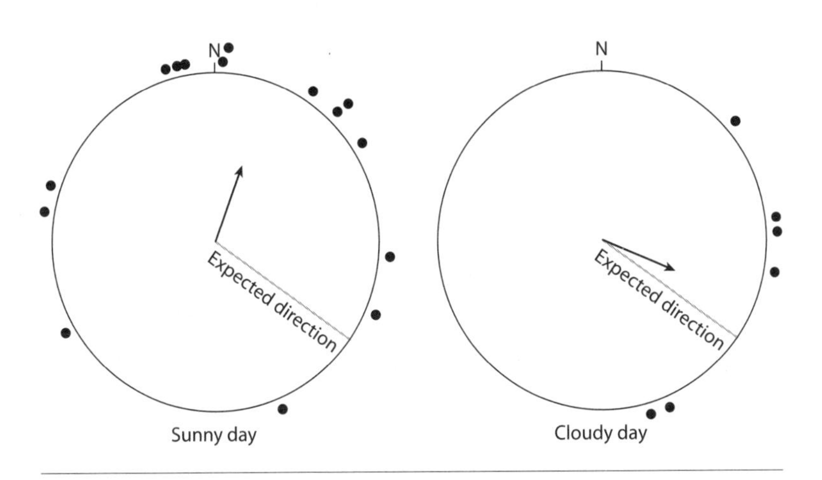

Effect of clock shift under sun and clouds. Under sunny conditions these clock-shifted fall-migrating sparrows chose a direction roughly 90° CCW of the preferred SE orientation of free-flying migrants. However, clock-shifted sparrows under overcast preferred the correct direction.

mal head) infinitesimal. An equally insubstantial and implausible contender 50 years ago was the earth's magnetic field, but this wild guess proved correct. A deceptively simple test by biologist William Keeton at Cornell changed the course of orientation research. Keeton compared pigeons wearing either magnets or equivalent brass weights under sun and overcast. Both groups were well oriented on sunny days. Brass-equipped birds did fine under overcast too, relying on the backup strategy evident in the clock-shift experiments. However, pigeons carrying magnets were unable to get their bearings when the sky was cloudy. The backup system, therefore, had to be magnetic. More remarkable still, while magnets do not affect sunny-sky orientation in birds older than three months, they disorient younger pigeons under all conditions. These tests and many others reveal that the inborn time-independent compass that pigeons rely on initially is magnetic.

A subsequent test was devised by Charles Walcott, an adventurous biologist who invented the radio-tracking technique for

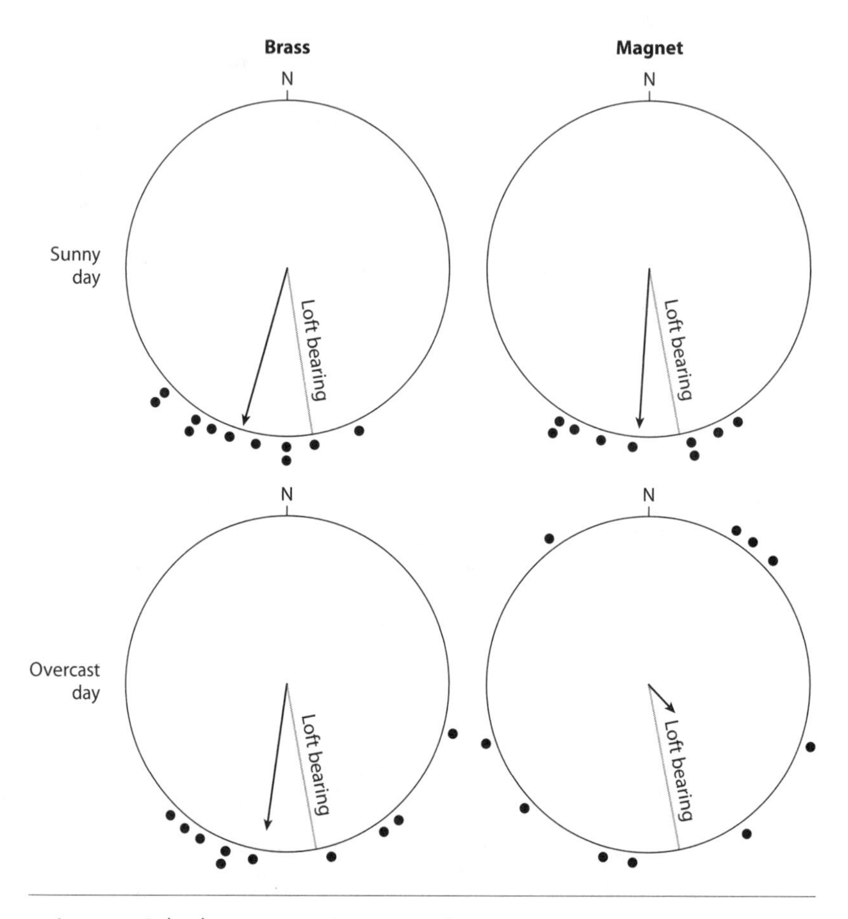

A magnetic backup compass in pigeons. Pigeons wearing magnets are disoriented on cloudy days, whereas control birds wearing equivalent brass weights are unaffected under overcast. Both groups do well under sunny conditions. Note that in all three tests with well-oriented birds the mean departure bearing is slightly CW of the loft direction. Each departure point has a characteristic release-site bias.

studying pigeons and later followed homers in his plane (documenting their unfailing proclivity to fly into restricted air spaces). Walcott mounted miniature, battery-powered Helmholtz coils, which generate easily manipulated magnetic fields, on the heads of pigeons. Control birds wore the same equipment but with the battery wire disconnected. When Walcott tested them under sunny versus cloudy conditions, turning the magnetic coils on or off on the experimental birds, an effect was seen with an artificial magnetic field only when the sky was overcast.

Working with Walcott, we discovered and localized a vast number of magnetite domains in pigeon heads in a small region known

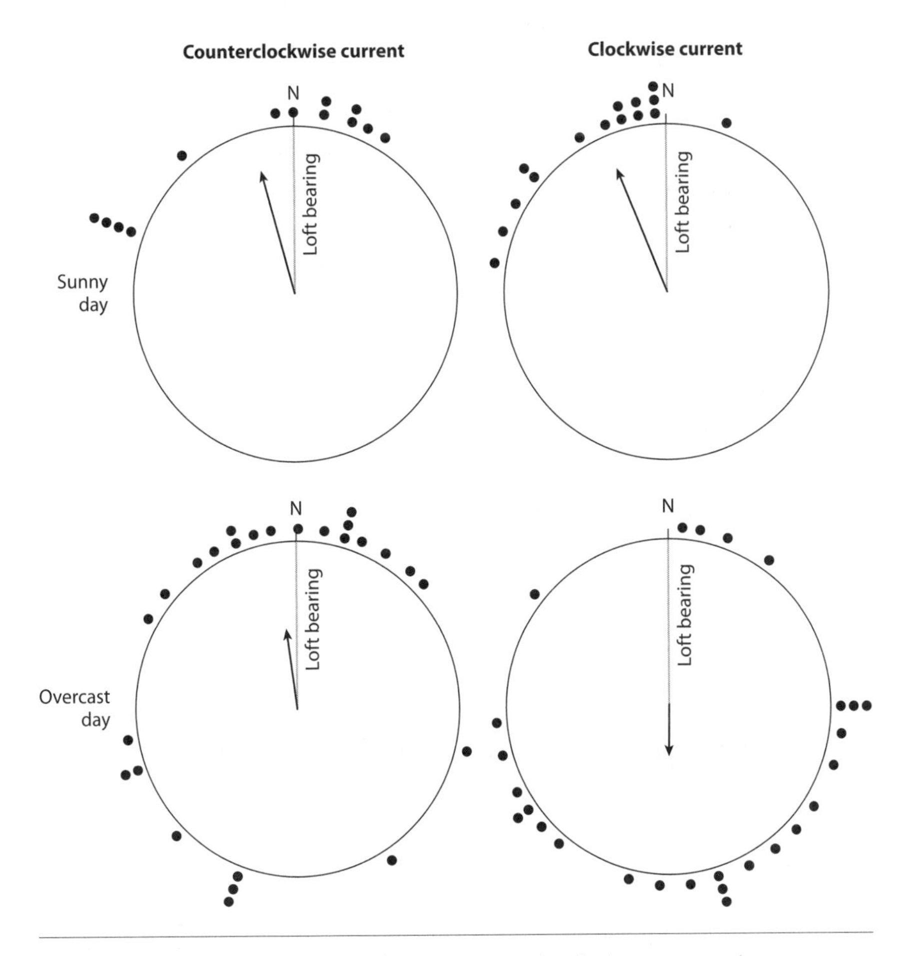

Coil-based tests. Pigeons wore battery-powered coils that generated a magnetic field inside and around the head. The coils had no effect on sunny days, but under overcast current flow that canceled or even reversed the field in the bird's head (lower right) reduced clustering and tended to send the animals away from their goal.

as the ethmoid sinus, located in the skull and beak between the olfactory and optic nerves en route to the brain. It seemed logical to assume that this was the structure underlying the magnetic compass sense. Other researchers then found seemingly identical organs in such widely dissimilar species as dolphins, salmon, and sea turtles. However, things did not resolve themselves simply.

Soon thereafter a methodological breakthrough allowed researchers working on caged European robins to reorient the birds' preferred hopping direction by altering the surrounding magnetic field. Exploiting this advance, Wolfgang and Roswitha Wiltschko of Goethe University in Frankfurt showed convincingly that only the field's dip angle mattered to these birds: a field that pointed down meant north regardless of its north–south polarity. Consequently, a field entirely in the horizontal plane—one that would rotate an artificial human compass into the standard north–south alignment—left the birds disoriented.

This is not the way a magnetite compass works. Subsequent tests by the Wiltschko group with pied flycatchers (birds that make a dogleg in their route) showed that if the dip angle was slowly altered to mimic southward movement toward the equator, and the timing of this change at least roughly matched the intervals of normal migratory flight, the animals would shift their hopping direction at the appropriate magnetic latitude for making their turn. But again, this reaction was independent of the actual polarity of the applied field.

Although the independence of the migrants' compass from polarity was a surprise, our group had shown early on in a theoretical paper that any paramagnetic or superparamagnetic compass would produce the kind of odd behavior the caged robins were showing. When physicists began suggesting that photosensitive pigments could respond paramagnetically, the possibility of a connection led fairly quickly to tests under various light conditions. Experiments by the Wiltschkos on low-light orientation with caged migrants quickly revealed that white, green, or blue illumination was essential for the system to work; red, yellow, and violet were ineffective. (White light is a mix of all wavelengths.)

As we saw in the previous chapter, experiments with pigment-based paramagnetism show that the axial response requires a minimum level of blue-green light—about 0.7 lux—to function. How much illumination is this? Bright starlight yields only 0.00005 lux. A three-quarters moon produces a useful 0.75 lux in the relevant part of the spectrum. These numbers present a potential problem: what is an animal en route to its winter or summer range to do on nights without sufficient light? For crepuscular migrants traveling just after dusk and just before dawn there is plenty of illumination. Even for determined nocturnal travelers this presumably is not a worry if the stars are visible to use as a compass (and under overcast fewer birds are aloft anyway). Nevertheless, to migrate in clouds on moonless nights birds probably need a different guidance system.

Another Wiltschko test recently revealed that in complete darkness European robins in a cage will adopt a NW orientation regardless of season. This behavior does not depend on a paramagnetic receptor: it does not require light, ignores the dip angle, and is not disrupted by the range of radio frequencies that disable the cryptochrome system. Instead the response is based on the permanently magnetic crystals of the ethmoid sinus. Somehow, when the cryptochrome approach is disabled, migrants in the wild evidently are able to use their permanent magnets as a kind of compass.

If we look at animals in general, a curious lack of apparent pattern emerges. Recall that one insect (the honey bee) has a light-independent compass, whereas another (the fruit fly) relies on a photopigment-based system. Looking at vertebrates, and moving up the phylogenetic hierarchy, we find that elasmobranchs (sharks and rays) may depend on induction. Salmon (the best-studied bony fish) and tuna have opted for permanent magnets in the ethmoid sinus. Newts (amphibians) require light, but sea turtles (reptiles) depend on magnetite. Among mammals, bats and the strange subterranean mole rat also use the permanent-magnet strategy. Some of these species also have a map sense (discussed in a later chapter), which is at the very least sensitive to magnetic stimuli. To the extent there is a pattern, the "typical" vertebrate species either has a light-independent magnetite-based compass only, a light-independent map sense and a light-independent compass system, or a light-independent map sense plus a light-dependent paramagnetic compass. Homing and migrating birds are in the latter group. In summary, the magnetite system looks ancestral, whereas the cryptochrome strategy appears more recently evolved.

What then does the coexistence of magnets and paramagnetic pigments in birds mean? The most widely accepted interpretation is that birds evolved a permanent magnet system first, then replaced it in part with a paramagnetic compass apparatus; hence the residual response of the magnets. This is not the only scenario, but it's the simplest and has many analogies with other cases of sensory replacement. But perhaps the permanent magnet system is not just a useless artifact; maybe the ability to orient reliably but "incorrectly" in complete darkness could mean that researchers have not yet found the right way to elicit a natural response from this system. Even if the fixed-direction orientation is what it seems, there is another very intriguing possibility: perhaps direction under low-light overcast conditions is computed indirectly via a multistep process.

Imagine that you are lost without a compass but have a beacon (perhaps a distant lighthouse) and a GPS. The GPS will give you latitude and longitude. Take a positional reading, then walk in a fixed direction toward the beacon and take a second map fix. From the two map points you can establish the direction of the beacon. This permits you to steer for home by setting off at the correct angle to the beacon—the fixed-angle strategy described in an earlier chapter. For the no-light test with robins, then, their beacon is to the NW; in the wild the birds would need to take their two internal GPS fixes, discover this axis, and then calculate the proper angle to fly. In the lab, of course, map position necessarily cannot change because the bird cannot go anywhere. Whether this is what robins do in the wild remains to be discovered.

■ Calibration and Redundancy

Homing pigeons have a somewhat less daunting task than migrants; they need not worry about nocturnal orientation and light levels because they steadfastly refuse to fly at night. Like honey bees, they have multiple guides—the sun, polarized light, magnetic fields, a local map based on familiar landmarks (which we will explore in chapter 6), and even a kind of GPS (examined in detail in chapter 7). Although the time-compensated celestial compasses of pigeons begin to take precedence while they are still juveniles, there is a gap of several weeks between fledging and mature (post–12-week) orientation during which the initial compass is magnetic. But magnetic north is a problematic navigational tool. As we saw in the previous chapter, it is not usually the same as geographic north; errors of up to 20° are common, and frequently larger in the popular Arctic breeding grounds. Over the limited flight range of foraging pigeons, however, there is no great chance that these changing declination values could interfere very much with navigation. Nor should the absolute error matter: a particular and systematic deviation in outward readings is exactly compensated for by the corresponding bias in return measurements. If pigeons were content to rely on magnetic north, problems would arise only with large experimental displacements of hundreds of miles, when these translocations are into regions with significantly different declinations.

The problem with declination to relative homebodies like honey bees and pigeons arises entirely because they inexplicably insist on converting to ever-shifting celestial coordinates—the sun and the polarized-light pattern it generates—that require a time-compensation system to decode. The sky and its polarization appear to rotate around the pole point, which lies on the true north–south axis. To use a magnetic compass as a backup rather than as a primary orientation guide, an animal must discover the celestial axis in one of the ways described earlier and then measure the declination. But if you are a short-range diurnal species, why go to all this trouble when a magnetic compass ought to be able to do it all?

There are two obvious possibilities. The magnetic compass might be less reliable or less accurate than a fully calibrated celestial compass. Alternatively, pigeons might until recently—in an evolutionary sense—have been long-distance migrators for which changing declination was a genuine threat. Both alternatives are correct.

The first hypothesis is easily tested: simply compare the accuracy of pigeons released under sunny skies with those sent out during overcast at the same sites. In fact, sunny-day tests yield on average a roughly 30% reduction in scatter compared to those run under clouds. The second theory requires merely a glance at the nearest relative of the pigeon. Recall that pigeons are domesticated rock doves; the closely related mourning dove, one of the most common American bird species, is migratory (though many overwinter as residents where bird feeders sustain them during cold weather). The extinct passenger pigeon, which once blotted out the sky with migrating flocks of a billion or more birds, was another close cousin. It's reasonable to suppose that homing pigeons are able to home because they retain the navigational machinery of their migrating ancestors, complete with its automatic switch to a time-compensated compass before risky long-range flying begins. Indeed young pigeons spontaneously begin flying to greater distances from the loft at about 12 weeks, just after the changeover from their initial compass and mapping strategies takes place.

But the pigeons are only a model for the navigation systems of true migrants. To discover if the compasses of migrants change with age or experience in an analogous way, researchers must rear chicks under controlled conditions until their first fall. Under these circumstances, if we cancel the earth's magnetic field but allow a view of the sky, young birds of most species use the pole point or twilight polarization in the fall to locate north, and then fly directly away from it. The default migratory direction, then, is due south. For wild-reared birds in their first fall, on the other hand, particular populations of various species each have their own favored direction—typically somewhere between SW and SE. Evidently some sort of fine tuning is missing in the experimental group.

Does the discrepancy result from the missing magnetic field, or from the inability of lab-reared birds to fly around locally in the

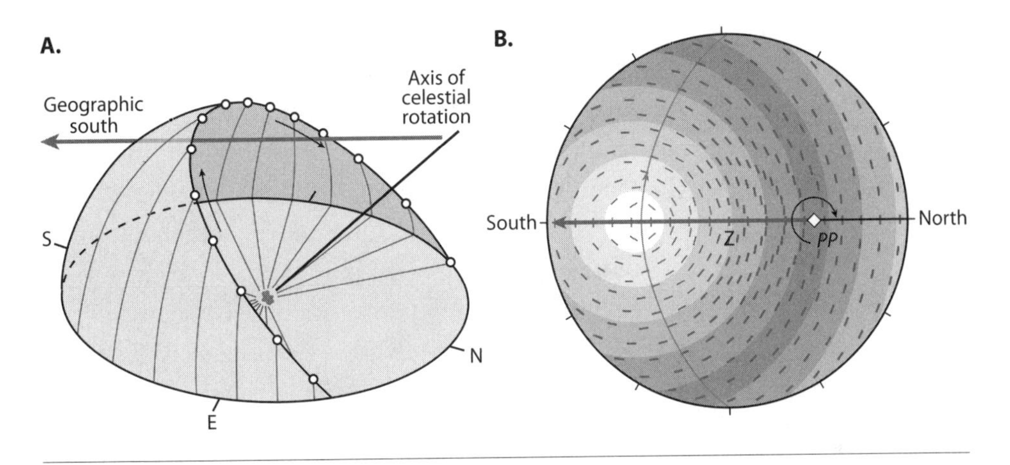

Default orientation. (A) An oblique view showing the axis of celestial rotation up through the pole point from the perspective of an observer on the ground, as well as the path of the sun through the sky. The sun appears to circle the pole point during the day, whereas at night the moon and stars will likewise seem to rotate about the same spot. (B) A fish-eye view from the ground looking up. The zenith is directly overhead; the rotational axis (pole point) is labeled *PP*. The pattern of polarized light in the sky appears to rotate around the pole point. Many Northern Hemisphere migrants know innately to fly away from the pole point whether during the day or the night. Of these, most also have innate, species-specific (often population-specific), preferred initial directions, specifying a particular compass direction to the right or left of due south.

breeding area? Add the magnetic field during testing, and the population-specific direction—SSW, say—emerges in birds with no prior experience in moving from summer to winter ranges. Remove the celestial cues instead of the earth's field, and the animals in the SSW population rely on their magnetic compass alone, and steer the same SSW taken up by their wild-reared colleagues. Apparently the birds enter the migratory period knowing to fly away from the pole point; they deviate from due south by an amount that is encoded into their DNA and measured by their magnetic compass. Only if you rob them of magnetic information do they fall back on the default bearing.

The innate headings that guide these first-time migrants can be inefficient at high latitudes where the magnetic declination may be quite large—90° or more in places. But as the birds move farther from the pole, the magnetic deviation from geographic direction declines dramatically, and their compass heading carries them close to their target. This route approximates a magnetic version of a straight Mercator track—what we called a rhumb line in earlier chapters. The birds know when they are close to their goal, probably by keeping track of latitude—or rather magnetic dip (inclination), which correlates with latitude fairly well. For species that do not follow their parents (nearly all species other than waterfowl) the youngsters must know innately which inclination signals the place to change direction or stop. But just as pigeons begin a preordained shift in their orientation strategy at a specific age, the first return journeys of many migrants, and all of their subsequent trips regardless of direction, are far more efficient than their first migration: they fly a great-circle route—the shortest distance between two points on the globe—and choose more precise initial headings. Natural selection has strongly favored anything that reduces the physiological stress of migration, as well as the increased exposure to predators that it brings. The basis of this shift lies largely in the use of map information, as we will see in a later chapter.

Just as with the pigeons, for a long time researchers knew a lot about orientation at the departure point (the breeding grounds, equivalent to the release site in homing tests), and the arrival site (the wintering area, or the loft for pigeons); what happened during the long and crucial in-between journey was largely a mystery. The first big step was to capture migrants at rest areas en route and test them in orientation cages. Because this approach allows the manipulation of orientation cues, it still offers many advantages over the more recent and dramatic technique of satellite tracking. We've already seen that for the typical trip north or south, lasting perhaps a few weeks in all, cues will change. Not only will declination progressively decline, but the sun's elevation, day length, pole point, polarization direction at sunset, and the constellations overhead will all shift. For a first-time migrant holding doggedly to a magnetic heading this may not matter too much, but for an older bird trying to plot the most efficient course, any systematic errors will lead it farther and farther off course.

Based on studies of migrants captured and tested in mid-journey, it's clear now that the key to dealing with this ever-increasing set of problems is recalibration en route. For instance, the pole point allows a bird to determine the magnetic declination, and thus to fly a true geographic heading in the absence of celestial information by using its compensated magnetic compass. This updated magnetic compass allows the migrants to calibrate the angle of polarization at sunset to allow for changes in day length and latitude. Now for a day or two at least they can use dusk polarization to calibrate their magnetic compasses. The updated magnetic compass can be used under the partly cloudy conditions so common during the migratory season to remap the constellations even if the pole point is obscured. The update on star patterns allows the animal to accurately place the axis of rotation. And so on, and so on, until the bird arrives at its target weeks later.

At first, this scenario for ongoing opportunistic compass calibration seems impossibly complex. The actual impossibility, however, was in the imagination of those of us who study orientation and migration. Imagination has been an even more limiting factor in understanding the next layer in the navigational puzzle. While compasses are critical for navigation whether local or global, in the air or at sea, their successful use very often depends on the animal having some sense of exactly where it is relative to its goal. We turn now to this question of position, looking first at short-range solutions, and then focusing on the most challenging mystery of all, the apparent true map sense seen in pigeons and many species of long-distance migrants.

Chapter 6

Piloting and Inertial Navigation

Wolfgang Köhler, a founder of the school of Gestalt psychology, is honored by animal behaviorists for his work on the mentality of apes. Given novel problems (a banana hung out of reach overhead, for instance), some of his chimpanzees could use tools such as sticks, boxes, poles, and the like to reach the food. Tellingly, they had to have played with the objects in the past before they could employ them successfully as tools under the pressure of the moment. Köhler saw in this an element of planning, an ability to use seemingly irrelevant information gathered in another context sometime in the past to solve a novel problem later.

The advent of the First World War marooned Köhler at his primate research station in the Canary Islands until the conclusion of hostilities. During this extended stay he made a huge if unintentional contribution to our understanding of navigation behavior. The subject of his analysis was his own dog, but some of the basic mechanisms he saw at work were similar to what was going on in the minds of his chimps.

Köhler noticed that his dog's capacity to solve a food-based navigational problem often depended on proximity: when it was close to a piece of meat the dog was unable to break its gaze and circumvent even simple obstacles between itself and the food. A short, isolated piece of fencing might as well have been a cage. When the dog's counterproductive fixation was interrupted or the animal was presented with the same challenge at a greater distance, the dog often saw the solution at once. For example, when Köhler tossed a piece of food out a window and then shut it, the dog stared out the window at the meat, pawing at the glass. When Köhler broke the spell by closing the shutter, blocking a direct view of the meat, the dog looked around and immediately ran away from the window, out the door, around the building, and to the food.

This ability only manifested itself if the dog was well acquainted with the room, just as the chimps could use tools only if they were familiar with them as toys. Both behaviors require what we now call *latent learning*—an ability to remember something irrelevant, an object or behavior that does not provide immediate reinforcement or have any evident utility. The dog solved the food problem by remembering that there was an indirect route to the goal, a path it had used previously. Subsequent tests showed that even having

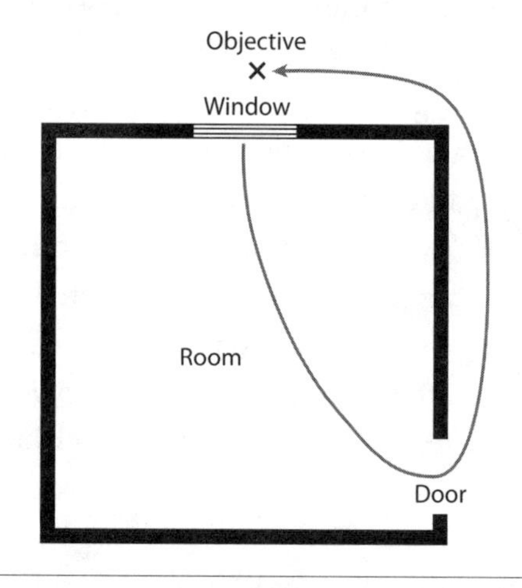

Köhler's indirect-route test. When the dog saw food tossed out a window, it attempted to get through after it. So long as the dog could see the meat it continued to struggle at the glass. When its view of the target was blocked by a shutter the animal turned and took an indirect route to the goal.

just *seen* the potential route previously could be enough for the animal to use it to retrieve the food—another apparent example of planning. But how is this mental map of the locale created, organized, and then employed?

■ Cognitive Maps

The reason Köhler's observations were such a shock at the time—so much so that many American psychologists did not fully accept them until the 1970s—is that they suggest that animals have a maplike picture of their surroundings, and use this mental picture to plan. Such an ability, commonplace among humans as we navigate around our homes, yards, stores, and places of work, seemed to be far too cognitively challenging for the so-called lower animals. The prevailing picture was an outgrowth of Behaviorism, the dominant psychological paradigm at the time. Behaviorism held that organisms are born with nothing more than a few reflexes, and accumulate what they need to know through chance associative learning. One form of this so-called *conditioning* involved learning to recognize novel cues and associate them with innately recognized stimuli, as when Pavlov trained dogs to connect the ringing of a bell with food. In time the dogs showed their mastery of this artificial relationship by salivating to the sound *before* the food actually appeared. The other type of conditioning involves learning a novel behavior though trial-and-error experimentation. The subject randomly tries various approaches, retaining behavioral units that bring it closer to its goal (obtaining a reward, avoiding punishment, etc.) and discarding those that do not.

According to the Behaviorist worldview animals can learn only through immediate reinforcement; moreover, they can learn only a limited range of stimuli. In the context of navigation a creature should be able to remember only locations and landmarks that provided practical rewards or memorable punishments—a list of spots characterized by food, danger, or mates for the most part. Understanding the relationship between two or more such locations would depend on opportunistic trial-and-error associations involving a reinforcement at one site followed by a reward at the second. To ethologists, the biologists who study animals in their natural environment, all this was obviously absurd. But under the controlled conditions of the lab, maze-running mice seemed to conform to these limiting strictures.

Köhler's dog (and chimps) clearly failed to fit the Behaviorist model. The dog solved the problem by realizing that there was an indirect route to the goal. According to Behaviorist dogma this would be possible only by painstakingly training the animal with rewards from the window, across the room, though the door, along the side of the building, and finally underneath the window. The real dog (as opposed to the theoretical Behaviorist canine) clearly understood the problem and its solution at an entirely different level; presumably it had some maplike picture of the room and surrounding area in its mind, and only needed to access it and use it to formulate a route to the meat.

The next psychologist to explore this ability was Edward Tolman, who was studying maze learning some two decades later. The most popular explanation of maze-running ability at the time was that the rats were learning a sequence of advances punctuated by turns—10 steps forward, a 90° turn to the right, 15 steps forward, and a 180° turn to the left, for instance. Hunger being a great motivator, training involved animals forcibly kept at 80% of their normal weight with food as a reward for performance. The rats were taught especially complex mazes by being placed near the goal and then restarted on subsequent trials a few steps farther away each time. Thus they built up a specific series of learned movements to get from the most recent starting point to the reward box.

Tolman was training rats in a well-lighted lab using a standard maze with its tunnels open at the top. He had occasion once to move the maze across the lab, and to his surprise discovered that his well-trained rodents were now lost. The same thing happened when he rotated a familiar maze: animals that had formerly performed flawlessly were suddenly clueless. Upon investigation, he found that his rats had been using landmarks on the ceiling to plot their course. Moreover, they knew enough about the layout of the maze that if he removed small sections of wall they would take shortcuts if the opening was in the direction of the food box, which they could not see directly. Rats do in fact also learn the motor sequence postulated by Behaviorism: when Tolman turned out the lights they were able to complete the maze without error, relying on a memory of the muscle movements learned during training. Moving or rotating a darkened maze does not interfere with their running.

So, did the rats have a map of the maze in their minds for use during the day? Tolman began investigating how rats come to learn about their locale. He provided opportunities for unreinforced exploration and latent learning—that is, empty mazes free of food. Without differential rewards a good Behaviorist rodent would not be able to learn any associations; it would leave the maze as ignorant as it entered it. The rats apparently have different ideas. In one memorable test Tolman provided a simple T-maze with empty boxes at the ends. One was dark and narrow, suiting the tastes of rats; the other was large and white. The animals explored the maze but, in the absence of any reward, should have learned nothing. Then on the next day Tolman took the same rats into another room, placing each one (in random order) into each of two similar boxes. The animals were given a mild shock while they were in the small, dark box—the one they normally preferred. Then on the third day the same rats were tested one at a time in the original maze. Each ran immediately to the end with the "safe" compartment. Since this maze-running behavior had not been deliberately trained, conventional conditioning was certainly not involved.

Tolman was the first to refer to the unreinforced acquisition of knowledge as latent learning. Köhler, as we saw, had already noted

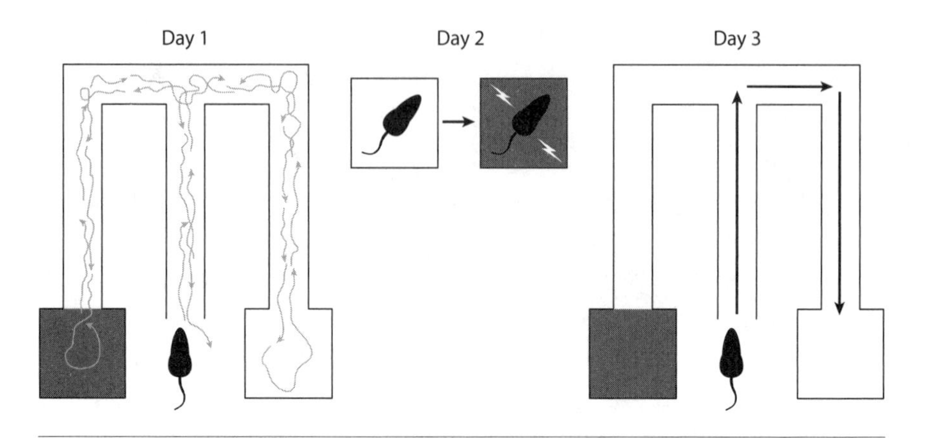

Latent learning in rats. The animals are allowed unreinforced exploration of the maze. On a subsequent day the animals are taken elsewhere and placed in boxes similar to those in the maze, and given a mild shock in one. On another day they are returned to the original maze, where they navigate immediately to the "safe" end.

that his chimps could only solve problems if they had previously played with and learned the potential uses of the toys. Tolman described the capacity of the rat to put together two seemingly unconnected experiences as a cognitive map. It's the rodent equivalent of being able to answer an exam question that is based on material from two separate lectures. While navigation was involved in most of Tolman's experiments he really meant "plan" rather than "map," but the spatial connotation has stuck, and is perfectly appropriate for our purposes.

Tolman's work, like Köhler's, was largely ignored; researchers were not ready to attach the term "cognitive" to anything but humans. Intriguing evidence continued to turn up, however. In one test, psychologist Lee Kavanau allowed well-fed (i.e., unmotivated) white-footed mice to explore an extremely complex 1400-foot maze. The animals discovered the shortest route (310 feet, with 1205 turns, avoiding the more than 700 feet of blind alleys) in under three days. Clearly this ought to be impossible.

Interest in Tolman's ideas was abruptly revived by the discoveries of Johns Hopkins psychologist David Olton in the 1970s. Olton devised a simple radial-arm maze with a release chamber that opened onto all eight passages. When a rat was set loose it explored all of the arms with, on average, no repetitions until each of the radiating options had been investigated. The route was not systematic, however; the rats seemed to be choosing the next passage at random. Olton thought they might be leaving an odor behind, solving the problem of avoiding repetition by choosing arms with no scent. But when he captured the rats after they had visited four passages and rotated the maze, they explored the four directions previously untried even if the reorientation of the maze meant going down alleys they had already investigated.

Just like Tolman's and Kavanau's animals, Olton's rodents were

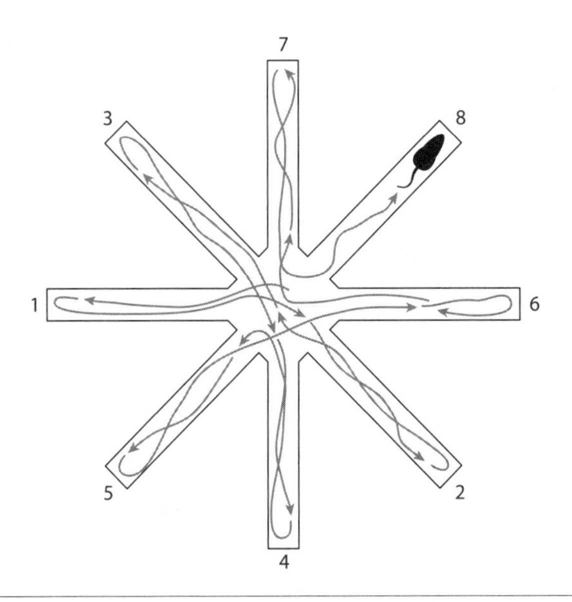

Rat running a radial-arm maze. The rodent explores the eight arms of the maze without visiting the same direction twice or using a systematic search strategy. The sequence here (arbitrarily taking north as up) is W, SE, NW, S, SW, E, N, and then NE.

keeping track of their routes despite the lack of reinforcement. Removed from the maze after exploring five arms, the rats would remember for days which three tunnels still needed to be checked. Later research determined that rats very sensibly remember which arms they have already visited until they have traversed half of the alternatives, then switch to remembering which ends they have not explored. A fairly consistent sex difference in this behavior has turned up: though both male and female rats solve the radial-arm challenge quickly and efficiently, they have different preferred strategies. For females, if visual cues ("landmarks") are available they remember the visited arms based on how they looked; males prefer to keep track of angles (i.e. the angle of each turn relative to the previous ones), thus solving the problem geometrically. Humans display similar sex-specific biases.

While definitions of cognitive maps vary from one researcher to another, we will use the strict Tolman sense, which invokes planning: the creature uses its understanding of the surroundings to form a novel plan. In practical experimental terms, this almost always means an ability to devise a new route from a familiar starting point to an equally familiar goal. In the simplest case, rodents that are able to exploit new shortcuts in a radial arm maze automatically demonstrate this cognitive map capability.

Following the discovery of this apparent mapping ability in rodents, a flood of studies showed that animals from fish to chimps create local-area maps, and use them to find the optimal route from one familiar place to another. Even certain insects and spiders have these home-range maps. We first noticed this capacity when we needed to train a group of bees across campus—a route that included a busy road, sidewalks with a steady stream of students, and inconveniently placed academic structures of one sort or another. Worse, our hive was on top of a building, which would entail lowering the feeder by stages down the outside wall. Reasoning that our foragers might have a local-area map we trained several to a feeder just outside of the hive and then captured them in a

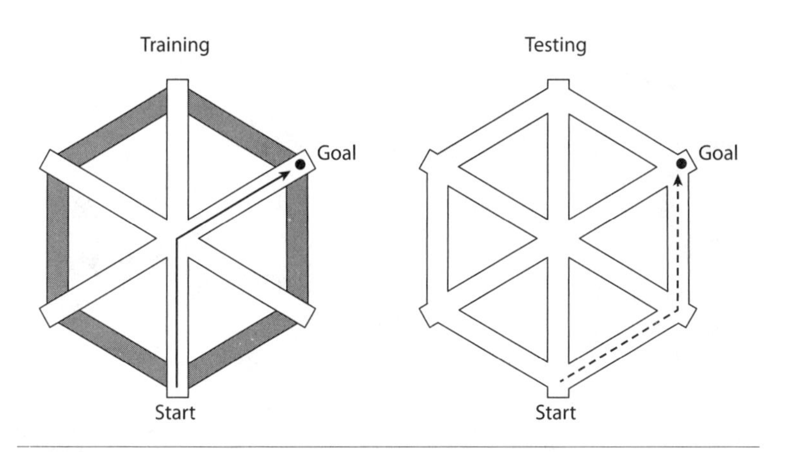

Route flexibility. Hamsters were allowed to explore a six-ended radial-arm maze, then were rewarded with food for traveling from the bottom tunnel to the end of the upper right passage. When the barriers blocking access to the circumferential route were removed, the animals opted for an alternative path even without being able to see the goal.

dark container and took them by motorbike to a barren parking lot 800 yards away to the ESE. After feeding the bees at this remote site we saw two sorts of behavior. Some circled repeatedly and then landed again on the feeder; others traced out a couple of circles and then flew off generally to the WNW. You can roughly judge the age of a bee by looking at her thorax and wings; as she grows older the fuzz on the thorax rubs off (the bee goes bald) and the wings become tattered. The bees that returned to the feeder had furry backs and unworn wings; only the more experienced bees took a shot at flying home.

The foragers that flew away from the feeder turned up on the comb a few minutes later and some danced immediately, signaling the parking lot. Apparently these experienced bees were able to deduce where they had been taken, presumably by recognizing the surrounding landmarks, and had then set course for the hive. We decided to look at this in a more controlled setting, using bees of known age and experience. A group of foragers was trained from

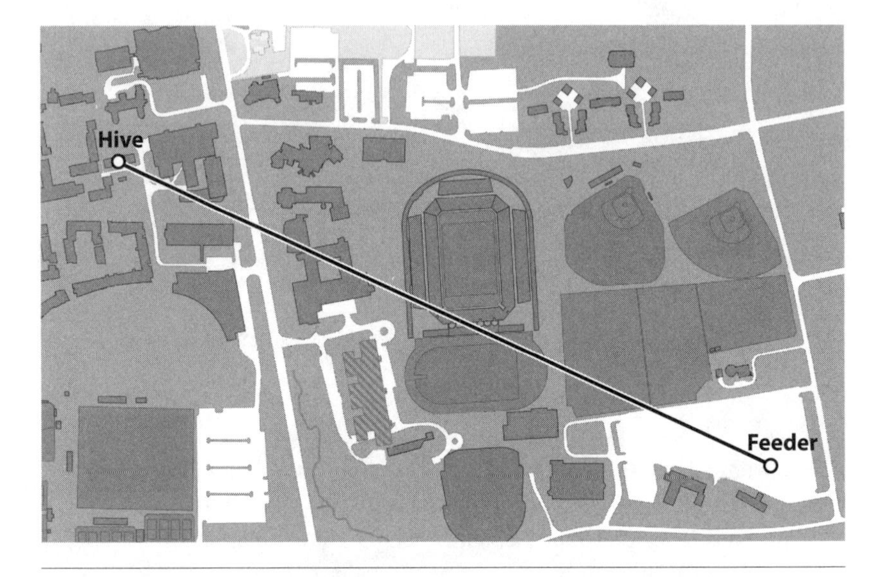

One-step "training." Foragers trained to a feeder next to the hive on top of a building were captured and taken in the dark to a parking lot 800 yards away. After feeding, some bees circled and returned to the colony, where their dances pointed 800 yards to the ESE.

their colony to a hidden clearing in a forest. The bees made repeated round-trips to the feeder for two hours in the morning over the next several days; presumably they spent the afternoons foraging in other places. For testing we captured the trained bees as they left the hive for the food station and carried them in darkness to another spot well within their home range but out of sight of the feeding station. The bees were released one by one. Each forager would circle, apparently looking at the nearby landmarks, and then set off along a direct route to the station to feed. Of course this behavior necessarily depends on prior familiarity with the local area, and the presence of large unambiguous landmarks so the bee can get its bearings.

We discovered that very young foragers were not reliable. Moreover, if we moved the release site farther from the hive along the forest edge away from a small but distinctive tree in the field, orientation also suffered. In short, this is exactly the behavior expected of an animal with a cognitive map.

Even more dramatic are tests with salticid spiders. These tiny hunters also are called jumping spiders, a reflection of their prey-capture technique. They often forage in dense shrubbery, tracking prey and seizing it in surprise attacks. Their telescopic eyes give them the best visual resolution of any terrestrial invertebrate, roughly equivalent to that of a small lizard. It must often happen that they spot a victim but have no direct route to it through the vegetation. In a maze in the laboratory they will scan the apparatus looking for the best way to get to the prey, even if it involves walking away from the food initially. As with bees and rats, chimps and dogs, the behavior depends on using available information to formulate a plan—a specific route in this case. In this species, however, no previous experience with the location is necessary; the spiders get all the information they need for this feat of restricted planning through real-time visual inspection.

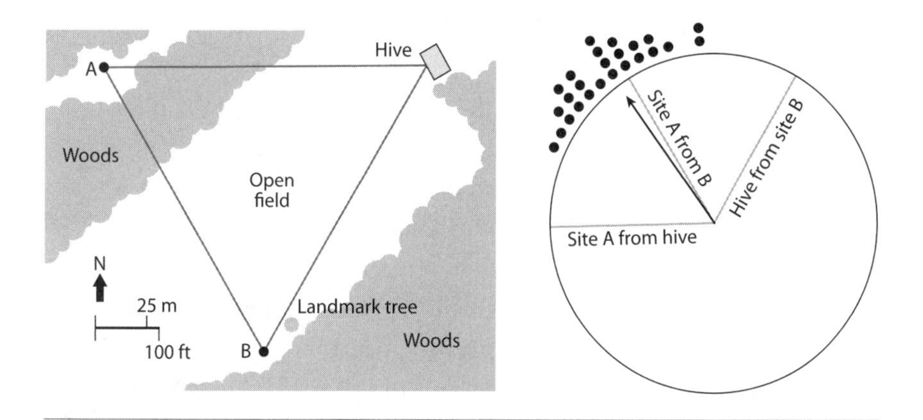

Kidnapped-bee experiment. (Left) Foragers visiting food station A were captured departing the hive and carried in darkness to release site B. (Right) When set free they circled and flew toward station A, hidden in the woods. Also shown is the direction predicted if the foragers had left the release site on the same bearing they adopted leaving the hive, and the orientation expected if they had simply flown back to the hive to get their bearings.

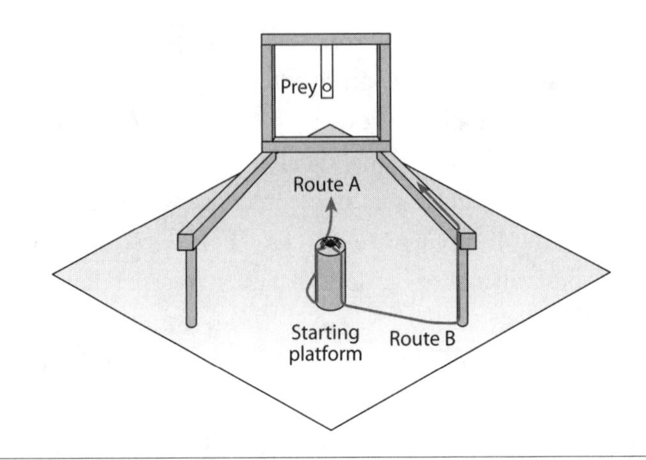

Route planning in spiders. The salticid is placed on the starting platform where it spots the struggling prey. The spider uses its high-resolution eyes to scan the surrounding area for a time, then sets off away from the prey to one of the two vertical posts. It climbs the post, checks to make sure the food item is still in place, and then sets out along the horizontal bar toward it.

■ Developing a Local Map

At first glance the distinction between dead reckoning and local area maps seems clear. The first involves collecting a list of abstract angles and distances on the way out and then making equally abstract computations to solve what is basically a problem in trigonometry. This is what desert ants were doing with such precision in an earlier chapter, using the angle of polarized light in the zenith and counting their steps. It's what the navigator on the *San Antonio* was attempting when drift defeated his efforts. We think of the geometry-rich process of inertial navigation as abstract because experience tells us (mistakenly) that humans only undertake this process out of dire necessity, after considerable training, and with specialized equipment. Local area piloting—learning the nearby landmarks and steering by them—is, on the other hand, something we do every day inside rooms or stores, between buildings, and en route to work or shopping. It seems natural and easy. Piloting may require experience, but surely not training, and definitely no trig.

If we think back to honey bees, however, the connection between these two strategies becomes clear. A new forager with her low-resolution vision sets off from the hive into a largely unfamiliar world. True, she has spent a couple of days performing short orientation flights, during which time she has memorized what the hive entrance looks like and learned the motion of the sun through the sky. But the moment she is even 50 feet from the colony she is out of visual touch with the home she was born into 15–20 days earlier. Of necessity this inexperienced insect must depend on inertial navigation to keep track of her position relative to the hive, recording the angles and distances traveled on each leg of her wanderings. When she is ready to return the forager computes her apparent position and, allowing for the intervening movement of the sun, sets course for home.

We know honey bees can do the necessary trigonometry for this dead reckoning with considerable precision because they can draw accurate maps—that is, generate the distance and direction components in their dances—after circuitous journeys through unfamiliar surroundings. We can quantify this capacity for at least a three-component trip. Von Frisch once moved a hive overnight to a new location next to a long multistory building. In the morning, of course, none of the landmarks were familiar to these displaced bees. He quickly trained a group of foragers around the end of the structure to the other side. When the foragers returned from a three-leg journey that led 275 feet 80° CW of the food, then 200 feet parallel to the hive–station axis, then 275 feet 80° CCW of the target (750 feet in all), they signaled a location only 275 feet away. This computed location was about 5° CW and 10 feet past the station they had visited along this indirect dogleg route. This corresponds to an error of less than 5%. Even with our much better visual resolution, humans walking the same course typically get the distance and direction wrong by more than 20%. Bees recruited by these dances, interestingly, flew up and over the building to reach the feeder, rather than around.

If its home range has useful landmarks, the forager begins to depend more and more on them to judge position—a sensible behavior because it reduces the chance of error from drift and compass ambiguity under overcast. Indeed bees will fly somewhat out of their way to tag a landmark en route to a familiar goal. But being able to use these isolated visual markers does not necessarily mean the bee has a cognitive map. There are two general ways animals can use landmarks, only one of which fits into our original orientation hierarchy. One is the cognitive locale-map technique. The other, the sequential strategy, does not require understanding the overall relationship between visual cues.

With this alternative approach an animal memorizes a sequence of "snapshots" of features encountered on the way to the

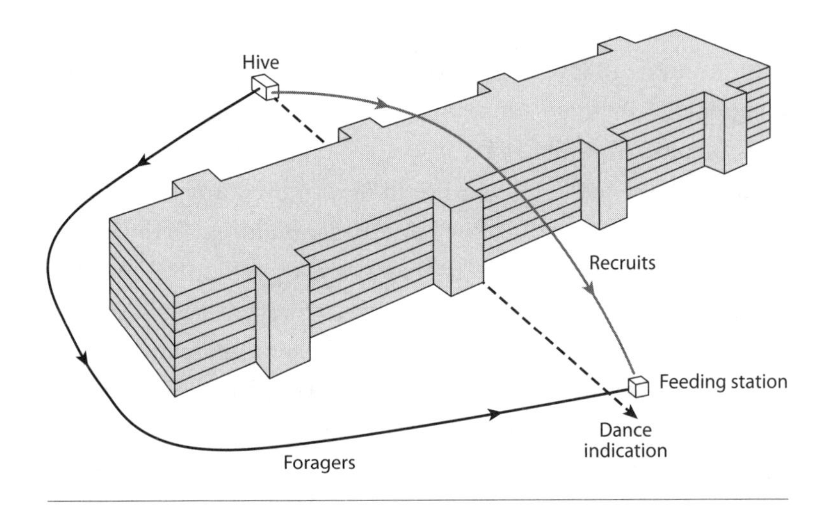

A honey bee detour experiment. Foragers were trained around the building along a highly indirect route. In the hive they danced to a spot within 30 feet of the true location. Their trigonometric integration of the three legs of the flight was accurate to within about 5%.

goal and follows this trail of visual cues out and back. Because this strip-map strategy seems simple, it was a popular model for insect piloting in the days when brain size was presumed to limit the navigational intelligence of insects. A lost bee, for instance, was imagined to fly about until it stumbled onto a familiar landmark. This allowed it to look for and fly to the next scene in its snapshot album, and then to the next, and finally back home. In fact strip maps (also called sketch maps) are a key step in forming genuine cognitive maps for some species, including humans.

Early in our own learning of a new locale we usually go through a stage in which we use landmark memory in a relatively inflexible route-linked way, unable to see shortcuts or to imagine what is on the other side of a building or hill. But with experience many navigating insects, like most birds and people, move on to the true local-area map, having learned enough about their surroundings to relate the positions of separately encountered landmarks to one another. There is some evidence that this integration is facilitated when two strip maps happen to intersect, sharing a landmark that binds them together. But though it may help, this overlap doesn't seem essential in at least some species. Initial landmarks, plotted on the basis of inertial reckoning, begin to fill some sort of mental grid, taking up positions relative to one another. This stage in the development of a plan is sometimes referred to as a bearing map; the original anchoring landmarks form a trigonometric skeleton awaiting the addition of finer visual details.

Once key landmarks and strip-map details have been combined, a kidnapped bee is able to set a unique course after displacement. A redeployed forager plots where it is based on adjacent landmarks, determines its distance and direction from the goal, and sets off along the correct bearing. The ability to fall back on the route-linked alternative under conditions of poor visibility, particularly after dark, provides a welcome backup. It's also a useful strategy when a journey becomes routine, allowing humans and presumably other animals to travel without thinking about the route.

One of the best illustrations in nature of how the locale map is structured and used comes from a remarkable symbiosis between ratels, or honey badgers, and the honeyguide, a bird with the rare ability to digest wax. When a honeyguide finds a beehive it positions itself between the hive and a ratel—or a human—and begins to call. If the ratel moves toward the hive the bird flies about 50 yards farther through the forest and calls again, guiding the helper to the hive. The bird waits while its accomplice tears the hive open and eats the honey, then takes the broken comb and its larvae as its commission.

At one time the birds were presumed to search at random for vulnerable hives after locating a ratel. Tracking, however, shows that they take a direct route from the spot where they find a helper to a honey bee colony they have selected. The bird knows the path, but does it have a cognitive map, or just linear lists of landmarked routes? Imagine trying to lead helpers on the basis of a flip-book of strip maps, where every badger you find must be along one of these familiar routes.

Researchers showed that the birds have local-area maps in two ways. First they let a bird lead them to a hive, but kept walking past. The honeyguide attempted to lure them back to the goal but eventually gave up and led them to another hive, and then another. Evidently the bird knew the locations of several hives, and how to get from one to another. The bird also was able to lead researchers to the same honey bee colony from different locations; it eventually led them along seven different routes before giving up. The honeyguide's behavior must depend on a mental map of the forest and its hives.

What do hungry honeyguides experience when they find a willing ratel? For a human a desire for food might call up a picture of a nearby snack bar, a mental map of its layout, and a glance in the direction of the shortest route. We can easily draw a local plan of the eatery or a map to show the way there. Moreover, most of us can read a map and, if familiar with the area being described, de-

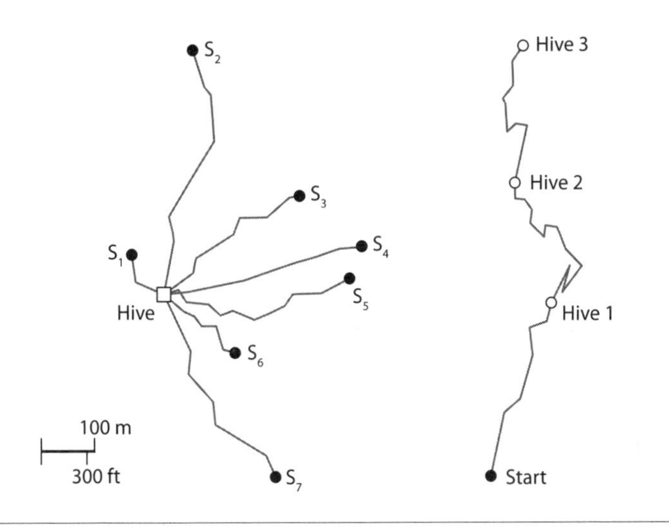

Honeyguide tracks. (Left) Indicator birds can lead humans through a forest to a honey bee hive from any location within about half a mile. (Right) The birds know the positions of several hives in their home range, and can lead a potential helper to one after another.

tect mistakes or incongruities. Presumably animals have the ability to envision layouts and routes as well, but what about a capacity to spot discrepancies? Surprisingly, this particular question seems to have been asked only of honey bees.

The issue is this: when a forager draws its dance-language map, can an attending recruit place the designated spot on its internalized cognitive map before exiting the hive? A graduate student in our Princeton group put this to the test by training foragers to a boat in a lake. The food he offered was suitably sweet and dancing was vigorous. But though many potential recruits attended, none came to the boat. Bees prefer not to fly over water, so perhaps that was the reason the dances were ineffective. But when the boat stopped near the far shore, recruits began arriving in substantial numbers. The most intriguing interpretation of these results is that recruits that attended the dances decoded the distance and direction cues, placed the location on their mental maps, discovered it was in the middle of the adjacent lake, and refused to act on the information. When the site being signaled was on or near the terra firma of the far shore, they were willing to investigate.

Being able to devise a novel route has always been taken as a cognitive achievement requiring mental visualization and planning. Our ability to draw maps, whether of rooms or campuses or towns or continents, seems to demonstrate a capacity for abstract thinking. But the data from bees and spiders should give us pause. Surely for them, this is a hard-wired process. But if so, why (other than romantic introspection) would we not assume that it is equally innate in birds and mammals? In fact, MRI and PET studies of rats show that map behavior depends on a small nucleus in the hippocampus, a long, curving structure moderately deep in the brain. (In nonmammalian vertebrates the equivalent structure is called the *pallium*.) Inside this module are at least three types of cells. One category, head-direction cells, encodes the direction the animal is facing. Grid cells divide the world up into a hexagonal matrix and plot where the individual is in its current hexagon.

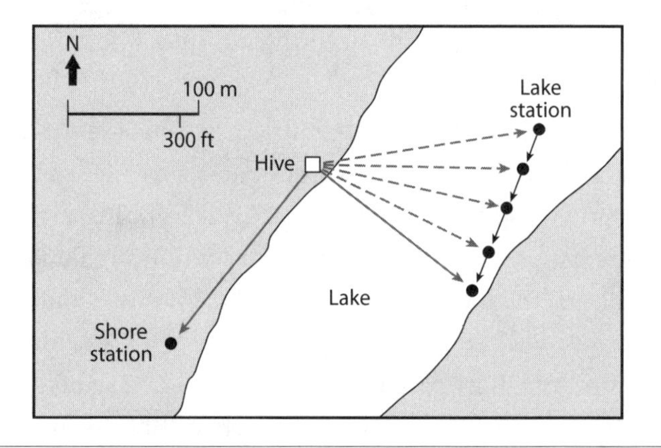

The lake test. A group of foragers was trained to a feeder in a boat. When the craft was in the most northerly position the dances failed to attract recruits. After the boat was moved south near the far shore, new bees began to appear.

(Different parts of the hippocampus have different grid sizes, ranging from about a foot to at least 10 feet, with much larger spacing at the end of the nucleus yet to be investigated.) Together these two bits of inertial information allow a third type of neuron, place cells, to fire when the animal is in a particular location in a familiar area. Lesions to this part of the hippocampus destroy the creature's ability to use a landmark-based map but leave its capacity to perform vector navigation intact.

The nucleus seems to be mainly concerned with processing; the specifics of landmark locations and appearance may be stored elsewhere. It seems to be wired in but empty at birth, and is filled in rapidly by exploration. Moreover, this mental drafting table has multiple place maps; the relevant one, populated by its set of learned landmarks, falls into place depending on where the animal finds itself. For a typical suburban human there would be one grid for each familiar supermarket, one for each frequently visited shopping center or mall, one for inside the home and another for work, one for the local neighborhood, others for larger-scale uses, and so on. If a human or an animal navigator is in an unfamiliar spot a blank grid is opened, ready to be filled in on the basis of new experiences. In short, there is no reason to think our mental maps are any more cognitively demanding than keeping our blood circulating or focusing our eyes on what we happen to be looking at.

■ Learning the Landmarks

The notion that map formation and use may be largely innate is reinforced by a number of studies of landmark learning. For instance, the use of landmarks is not a matter of simply taking a wide-angle photo and then trying to match the image later. As early as the 1920s Niko Tinbergen studied landmark learning in digger wasps, hunting wasps that prepare burrows to receive the paralyzed prey that will sustain their offspring. These insects must

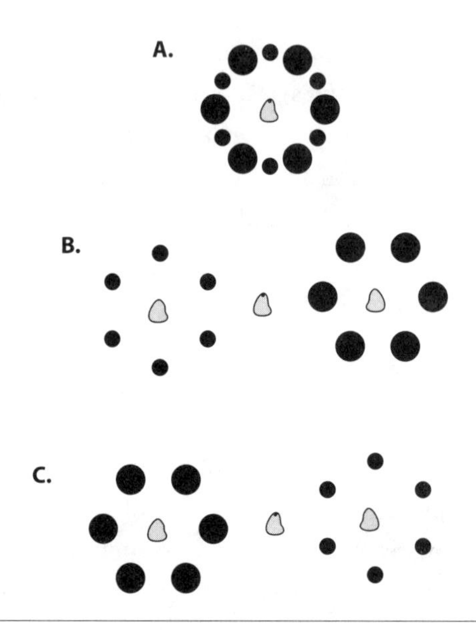

Landmark learning in wasps. Tinbergen surrounded a wasp burrow in the process of being built with alternating landmarks. While the insect was away hunting he created a pair of false burrows, each surrounded by one of the two landmark types. The choice by the returning wasp indicated which markers were more salient.

be able to find their nests from hunting grounds hundreds of feet away. Depending on the landscape, they rely on large-scale landmarks or dead reckoning to get near the nest. But to zero in on the tiny entrance they must memorize the location of the opening on the basis of nearby landmarks.

As a wasp was excavating her nursery in preparation for provisioning, Tinbergen would surround the burrow with markers of varying shapes and colors. The wasp would circle and study the cues before setting off to hunt. While she was away Tinbergen would create two artificial nests, each within an array of one kind of marker. The returning wasp's choice indicated which landmarks she was using to locate her nest. Tinbergen's wasps strongly preferred three-dimensional markers: the wasps ignored a flat disk in favor of a much smaller sphere. Dark landmarks were more popular than light-colored ones. Vertical objects were far more effective than the same markers lying flat.

Highly systematic follow-up experiments have since been directed at honey bees. Foragers learn the landmarks surrounding the flowers they are exploiting, most likely to compensate for the poor visual resolution of their compound eyes. Though bees appear to study the flowers during approach and again after takeoff, the two phases are quite separate learning exercises. The incoming bee is learning the flower's odor, color, and shape; when it leaves it is committing the surrounding landscape to memory. Demonstrating this processing dichotomy is straightforward: present one set of landmarks on arrival, and substitute a second while the forager is feeding. Let the bee observe these new ones on departure, then offer a choice during its next visit. It will opt for the markers it observed as it left. This is not a matter of preferring the most recently viewed cues. If instead of substituting a new set of objects before departure we just remove the cues it saw on arrival, the returning bee exhibits no preference for the initial stimuli.

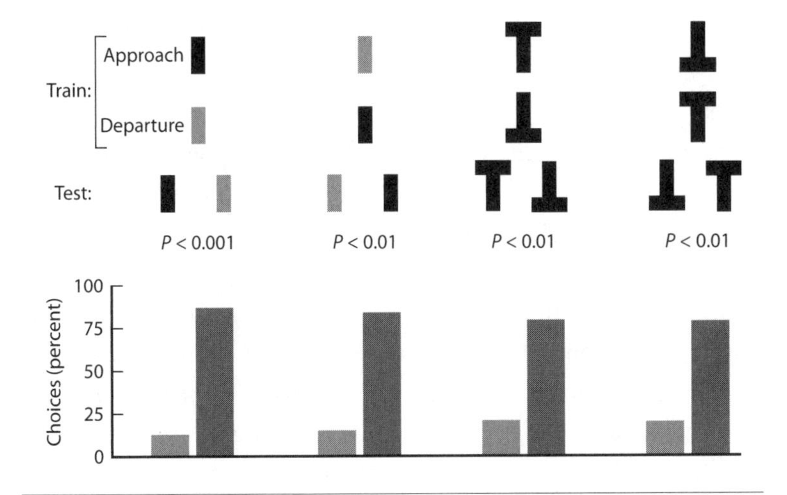

Timing of landmark learning. Objects seen on departure are remembered in preference to ones experienced on arrival. In contrast, the bee learns the odor, color, and shape of the food source during the arrival phase.

When we give the bees alternative landmarks to learn during departure and choices upon return, we find as did Tinbergen that three-dimensionality and texture are more readily remembered—or rather, more reliably chosen on subsequent visits. The bees can learn both the landmark's shape and color, but shape is more important. Foragers also learn something about the relative distances of the cues, probably from the parallax movement induced by their circling; nearby objects shift position more than do distant ones.

But easily the most important factor for foragers is angular position relative to the goal. If the size and shape of the markers are changed but the flying bee can find a perspective from which the relative angles are what it learned originally, it will pick such a configuration over an alternative with the correct objects in unfamiliar positions. If a few of the landmarks are removed after training the returning forager does not necessarily match its memory to the new reality in the way humans might. It instead tries to interpret the surviving objects as being the most widely separated of the original markers—a strategy that maximizes the precision of the bee's triangulation. Add a strong compass cue, and the rules for weighting the relative angles of landmarks versus their appearance changes. A failure to understand this innate situational hierarchy led early researchers to conclude that bees either cannot learn anything at all about their surroundings, or simply get it wrong. Rats given similar cue conflicts also opt for the interpretation that places the cues farther apart, so this may be a general contingency plan in animal orientation.

There is no doubt that bees and wasps are born already knowing something about how to choose and use landmarks. Even the remarkable desert ants that can locate their nest entrance in a featureless habitat through precise dead reckoning will seize upon the slightest landmark near the opening as an efficient guide for their final approach. The same innate preparation is seen in many birds. Certain food-caching species can learn hundreds or even thousands of locations where they have stored seeds for the winter.

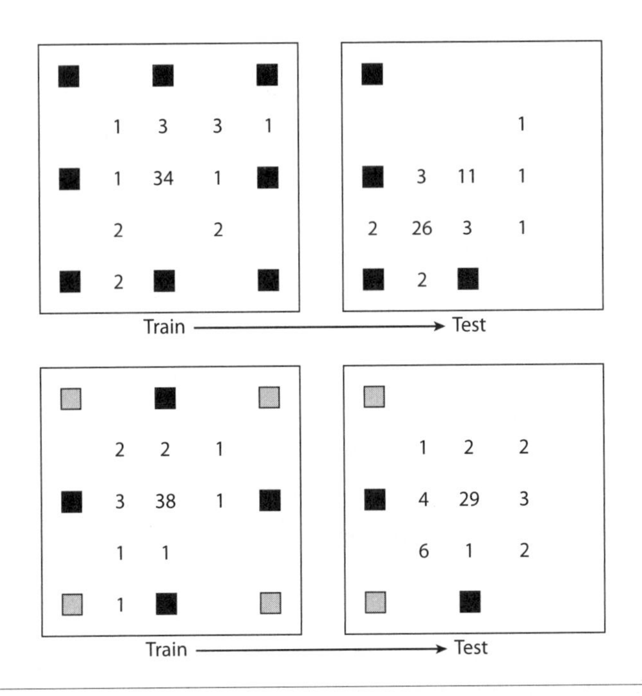

Importance of landmark configuration to bees. Foragers are trained to an inconspicuous food source surrounded by an array of landmarks. Two examples are shown on the left, one with identical cues and the other with two different kinds of markers. The numbers indicate the number of landings during periods with the food gone; the bees explore ever-wider areas in an attempt to find the missing sugar water. In the testing phase some of the landmarks are removed. When the markers are all identical the bees search closer to the remaining cues (upper right). This is an attempt to put the surviving objects as far apart visually as possible. However when the markers are distinctive (for instance, have dramatically different colors) the bees notice the discrepancy and land in the original location.

Clark's nutcrackers, for instance, remember each of the 6000 places they have hidden their winter supply of 30,000 pine nuts. The birds are careful to cache near landmarks, but not too near; thieving mammals and birds focus their searches around landscape features as well. And yet, subject these birds to a clock shift that puts the celestial compass into conflict with the intricate network of landmarks they have so painstakingly learned, and the birds are completely disoriented. At some level they trust the sun or polarized light as more reliable guides than transitory landmarks such as piles of leaves, fallen branches, and bushes.

Release a pigeon 60 times at the same site, with plenty of opportunity to learn the array of surrounding buildings and hills, and it too will still be fooled by a clock shift, even if the site is well within its daily range of exercise flights. Yet in the air a pigeon as much as four miles from the loft will drop into a stereotyped strip-map–based final approach, ignoring celestial cues. As with bees, which display the same habitual end route to a familiar source though on a much smaller scale, the navigational rules written into the onboard guidance modules of pigeons take precedence over human logic.

■ Local Navigation by Humans

As we've seen, high-precision local navigation in many species is the result of an innately guided process that begins with mapping landmarks inertially, relying on one or more compasses and an ability to measure distance. In most animals this leads to the development of strip maps. In time the hippocampus or its equivalent combines the strips to generate a true cognitive map, and with it an apparent understanding of the habitat and an ability to formulate novel routes. What about our species?

In an unfamiliar area, or even a well-known region as the sky goes dark, we imagine ourselves to be helpless. And yet there is good evidence that our species is not entirely naïve when it comes to navigation, and moreover we get better at using mental maps as we get older. Let's start with the worst case. We stand blindfolded in the open in the absence of any useful beacons—no sun warming one side of our face, no sound in the distance, no breeze. The result is that we walk in circles. Some individuals meander in relatively tight loops, whereas others trace more gradual arcs. Though no one walks a straight line, we all think we are maintaining a constant heading. Without a beacon or compass, humans trying to orient in or map their surroundings are in trouble.

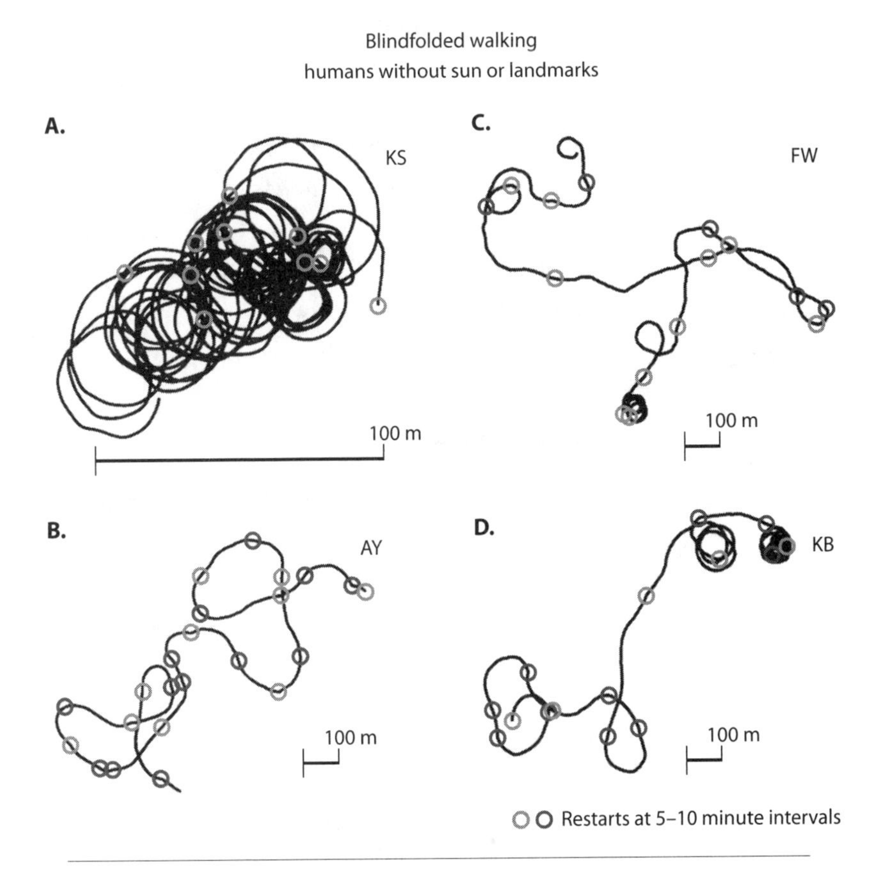

Walking in circles. Four individual humans were blindfolded and asked to walk straight (without sun or landmarks). Marks on the routes represent short pauses at intervals of 5–10 minutes imposed by the experimenter. Note that circles of the individual in A are relatively tight; the scale is expanded eight times compared to B, C, and D.

When we first began working on bee and pigeon navigation we were struck by our ignorance of the sun's movement and its potential use in orientation. Asked which way is north, most of us think first of street bearings rather than where the sun is likely to be at this time of day. But apparently humans are perfectly capable of using the sun as a beacon. For example, most unblindfolded humans asked to walk in straight lines across a featureless desert were able to keep a consistent bearing over distances of a few miles, though the direction walked was typically wrong by about 45° over much of the track. It seems that we use the sun unconsciously, though fairly incompetently, as a beacon. As a species that evolved in the rain forests of Africa, perhaps this should not be a surprise, though even in the forest a primate needs to be able to walk a straight line. For the !Kung bushmen of the Kalahari Desert, on the other hand, the sun's position in the sky is a key piece of learned and practiced life-or-death informa-

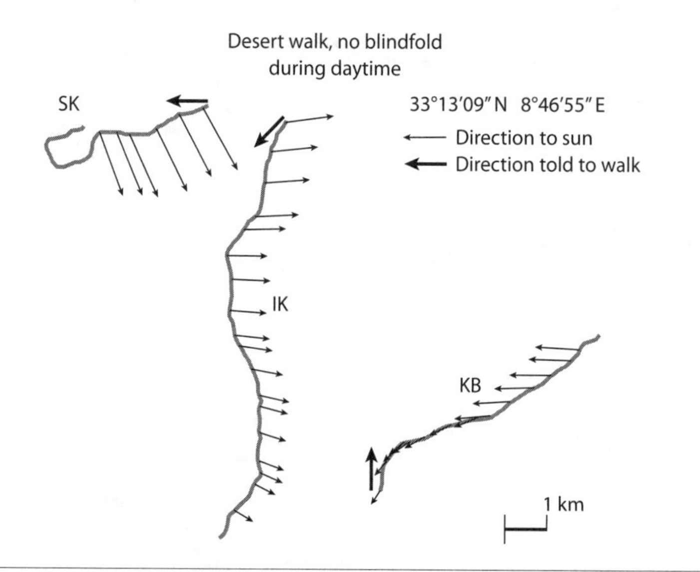

Walking in the desert. Unblindfolded humans walking in the desert on a clear day do not usually circle, but their tracks do diverge from the initial direction they are asked to maintain.

tion as they navigate over hundreds of square miles of relatively empty terrain.

Suburban-bred humans are, in fact, at their best in forests, at least when the sun is shining. In the absence of a solar beacon humans trace out wandering routes, usually crossing their own path multiple times. But if the sun is visible through the canopy the routes are straight and true for miles on end—and as far as one can judge from this tiny sample, the travelers even compensate for the sun's westward movement. Why is this so much better than in the desert, where the sun should have been far more visible and easily used?

The forest is an environment with a series of short-range beacons, something like the frog ponds birds may use under overcast at night. The walkers seemed to take their occasional views of the sun as a compass update, aim for a tree in the appropriate direction, and home in on it until the next unconscious compass reading. Returning pigeons do something similar, steering by beacons on the ground such as roads to provide an easy temporary guidepost. If the sun's movement is actually being considered, this piece of sophisticated but unwitting navigational machinery may be a part of our human inheritance.

So perhaps we, like the bees and pigeons, have an unconscious sun-compass capacity. As we saw in Chapter 2 we also have an innate ability to judge distance. This is based on two inputs. We use the semicircular canals of the inner ear to gauge movement, unconsciously integrating acceleration to get speed and angle of turns and then integrating speed to get one estimate of distance. As you will recall this computation requires judging time intervals, something we are not very good at. At the same time we unconsciously count steps and infer distance traveled from this input. As adults we weight and combine the two estimates to give us our dead-reckoning sense of where we are.

From this relatively imprecise bit of inertial mapping we can place landmarks in space, incorporate them into our strip maps,

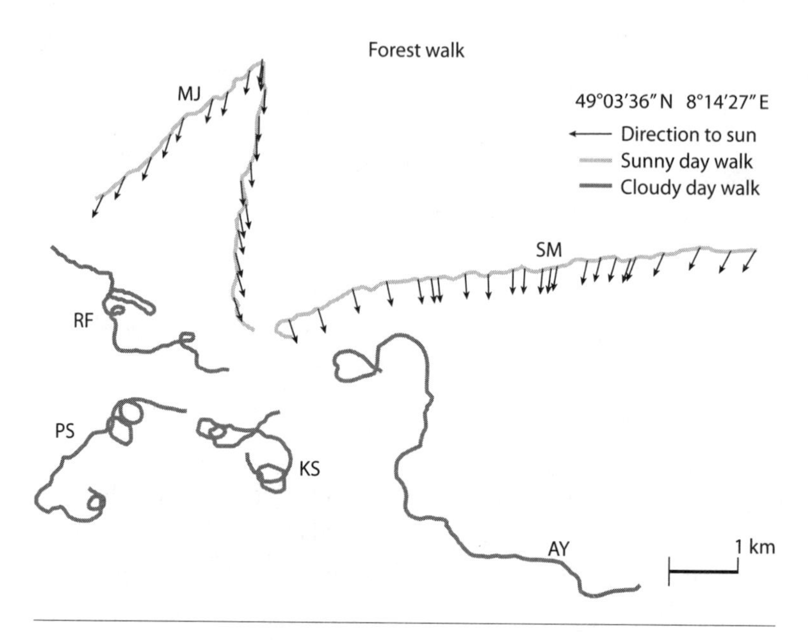

Walking in the forest. Four human tracks (RF, PS, KS, and AY) are shown for a cloudy day. The tracks wander, sometimes in relatively tight circles. On sunny days the three tracks (MJ, tested twice from the same starting point, and SM) are surprisingly straight for miles, even showing intriguing signs of sun compensation.

and then use the intersecting sketches to create hippocampal grids. As with migrating birds, there is clearly much opportunity for calibration. Multiple estimates of the distance and direction to a landmark will lead to its more precise mapping; the presence of a well-known landmark can help calibrate a later inertial calculation. Some ongoing calibration is essential as we grow when, among other complications, our stride length gets longer (and, ultimately, shorter).

Just as the orientation strategy of bees and pigeons changes with age and experience, so too does human technique. Several methods have evolved to examine this. One of the most informative is to test children and adults in a large, dimly lit room with a few readily identifiable targets on the floor and visual cues (aka landmarks) on the walls. The task can involve simple visual observation or active movement on the part of the individual being tested. The first kind of task depends entirely on remembering visual geometry and inferring from that the relative locations of what can be seen. The second task adds inertial data about distances and directions generated by the person's own movements. For instance, after an object on the floor is passively illumined, the subject is blindfolded. The markers on the wall are shifted and the lighted object is removed. Then with the blindfold off the subject is asked to point to where she thinks the target should be. Alternatively, the subject might be asked to carry the target to a convenient spot, providing herself with unconscious inertial information on distances. Then with the subject blindfolded the objects are moved as before, and the person is tested. Any improvement would come from the inertial data generated by the individual's movement.

In a typical experiment, the total errors in positioning are age related: four- to five-year-olds are usually more than 30 inches off target, while seven- or eight-year-olds are about two feet in error; adults miss by only about 13 inches. But more to the point, children opt for inertial information or geometric estimation, choosing between them seemingly at random. Adults, on the other hand, use a sophisticated weighted average of the two sources of data. We remain blissfully unaware of this automatic measuring, processing, and recalibration. The progression from bearing maps to strip maps to cognitive maps, a process discovered in insects, birds, and rodents long before the human parallels became clear, is beyond our perception, and almost beyond comprehension.

While the puzzle of local area maps and their neural underpinnings is beginning to be understood, it's not hard to see how members of our species have been driven to scale up personal locale maps to encompass the planet. Like Kavanau's white-footed mice, we are compelled internally to explore and fill in details even (or perhaps especially) in the absence of need or overt reward. But the human drive to map operates in the absence of a biological ability to know our location outside a familiar home range. For many nonhuman migrants a large-scale (potentially global) positioning system seems to be part of the onboard equipment. How it works is still a matter of considerable debate. The map sense is the final challenge in comprehending the mystery of true navigation.

Chapter 7

The Map Sense

The Arctic hosts immense populations of breeding birds exploiting the long days and the brief but astonishing productivity that the extended sunlight of summer makes possible. Two species of large shorebirds illustrate some of the extremes in bird migration. With wingspans of about 2.5 feet, these birds are large enough to carry transmitters, so researchers have been able to track their peregrinations.

The bar-tailed godwit nests on the Arctic coast and tundra, probing the mud and marshy ground with its long bill for mussels and worms. There are perhaps 100,000 bar-tails in Alaska alone. They abandon their young before fledging, leaving them to migrate south on their own. Heroic satellite tracking by Robert Gill of the US Geological Survey and Brian McCaffery of the US Fish & Wildlife Service reveals that the godwits pursue basically straight SSW routes all the way to New Zealand. A few birds truncate their journeys and overwinter in Melanesia; others pause there and then complete their trip; yet others arrive in New Zealand after 7000 miles of nonstop travel. Do the fledglings, flying south a few weeks later, know where they are going, or do they just fly a compass bearing until (if lucky) they happen to spot land in the barren vastness of the Pacific?

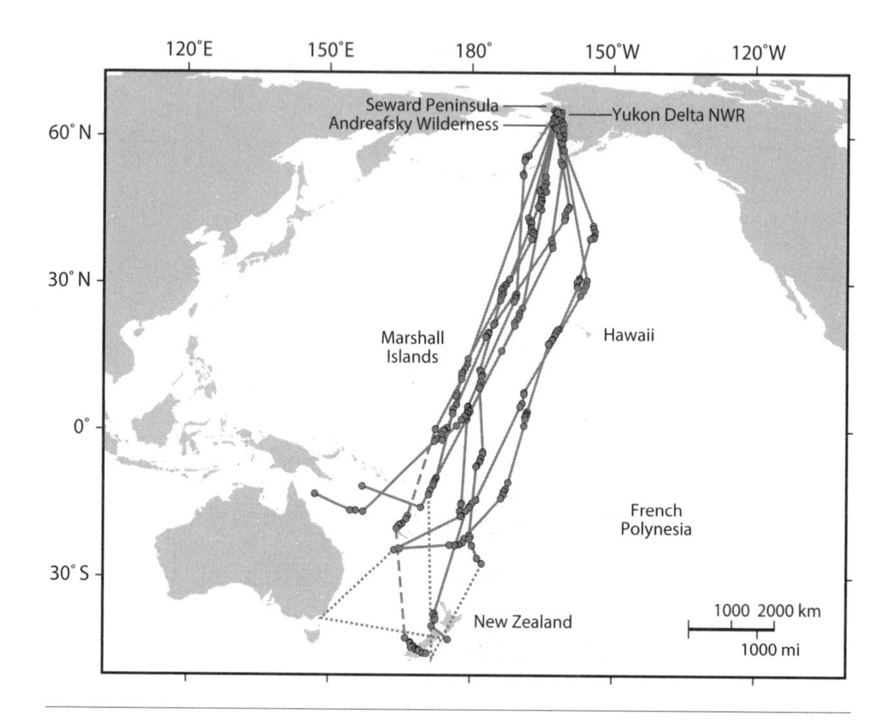

Migration tracks of bar-tailed godwits. Birds from the Yukon Delta National Wildlife Refuge in Alaska depart SSW and fly as much as 7000 miles nonstop to wintering grounds in the South Pacific. Some arrive in New Zealand directly from Alaska, while others make brief stops in Melanesia. Nonstop flights are shown as solid lines; dotted lines indicate routes after first landing.

The answer may lie with the other well-studied Alaskan migrant, the bristle-thighed curlew. These shorebirds are slightly larger and far less common than the godwits, with a population of perhaps 7000 in all. They make their homes in one of two small lichen-dominated shrublands overlooking the surrounding tundra in Alaska, eating mainly berries. Like the godwits, the curlews abandon their young before fledging, leaving them to find their own way; thus any information about goals or bearings must be innate.

For years researchers had noticed that a few of the curlews banded in Alaska's Andreafsky Wilderness either wintered in Hawaii or made brief stops there. None of the much larger cohort tagged on the other breeding grounds (the central Seward Peninsula) were ever seen in Hawaii. This suggested the two populations might take different routes, and perhaps have different wintering grounds. In 2006 some of the Seward curlews were tracked. Unlike the godwits, they departed SE and flew to within 600 miles of California before turning south and arriving in French Polynesia after a nonstop journey of just over 6000 miles. Their dogleg route came nowhere near Hawaii; no wonder they've never been seen there. The next year birds from the Andreafsky Wilderness were tracked in their turn; they flew south over Hawaii and then turned sharply SW, arriving after 5300 miles of continuous flight in the Marshall Islands. The juveniles that followed a few weeks later doubtless utilize the same group-specific route.

How do the curlews manage to find these small targets? We've looked at the compass mechanisms that help orient birds; perhaps they fly a predetermined set of bearings—that is, they might rely on vector navigation, shifting direction at predetermined latitudes, and have no real map sense at all. When the target is something the size of South America or sub-Saharan Africa, this seems a very reasonable hypothesis. But if the goal is New Zealand, French Polynesia, or the Marshalls we must wonder what fraction of the chicks could hope to stumble on these islands after 6000 miles of windblown travel over open water. For them a vector strategy would certainly be sufficient during the first three quarters of the trip; but given that there is a lot of ocean and very little land, a map would be helpful toward the end. When monarch butterflies travel 2000 miles or more from their birth sites in the northeastern United States to a particular mountain in Mexico, vectors might get them within a few hundred miles, but some sort of map is surely necessary to account for their observed precision. In the ocean equally substantial migrations are seen in salmon, eels, and sea turtles; they arrive at the exact river mouth or natal beach with the precision of ballistic missiles.

Most researchers agree that many species of long-distance migrants must have a map sense; the animals know at least roughly where they are relative to their goal even if they are just flying or swimming a vector for the moment. They cannot afford to depend on vectors and serendipity. What scientists cannot agree on is exactly what the map sense does, or how it works, or even whether there is just one mechanism. Whatever processing the animals are doing, the result can be relatively precise; as we saw, the pigeon map seems accurate to about a mile, at least near the loft.

■ Do Animals Really Have Global Maps?

There are at least two good reasons to have a map. One is relevant for first-time migrants and birds that use a vector strategy initially—animals that behave as though they don't know or care about their exact location en route. A map allows an animal to find its target once it is near the goal, perhaps as it senses that it is approaching the appropriate latitude. The other less obvious reason is that only a map allows an animal to correct for en route mistakes, such as an undetected crosswind or an inaccurately measured solar azimuth. An early vector-use error tends to magnify itself; a map-based midcourse correction puts the creature back on track. For wide-ranging navigators a map is not really a luxury.

We are accustomed to think in terms of a coordinate-based map. For humans, each point has an east–west longitude and a north–south latitude. Thus if we are in Princeton (latitude 40°21′, longitude 74°40′) and wish to travel to Boston (latitude 42°21′, longitude 71°4′), we can calculate that we need to travel NE (52° CW of north, to be exact) for roughly 235 miles. This is well within the 600 miles/day range of homing pigeons. But is this what other animals actually know about their world?

Possibly because we have been indoctrinated by two-dimensional maps, and are used to words and tables being laid out in a grid, we think it natural to use start and end points separately defined along orthogonal coordinates, and then to convert them into a radial (distance and direction) result. It's tempting to suppose animals do the same. But some or all species might instead depend on radial coordinates throughout, as honey bees seem to do.

Whatever the strategy, any hypothesis must account for the consistent but counterintuitive observation that at least in homing pigeons, the precision of their orientation toward home does not decline with increasing distance. To be taken seriously any hypothetical map strategy must predict that animals know where they are with the same resolution regardless of how remote they are, at least up to several hundred miles. Keeping track of displacement inertially, on the other hand, should lead to less and less accuracy with distance.

Another factor to keep in mind is that map and compass use may change with age and experience. We have already seen that pigeons imprint on the loft location just after fledging, but only if they are permitted at least short exercise flights in the vicinity. Under normal conditions birds make two other major shifts after leaving the nest, usually between their fifth and twelfth weeks. First, young pigeons gradually move from a nearly complete dependence on their magnetic compass to a preference for solar cues—a change that depends on learning about the sun's movement through the sky (or the corresponding shift in polarization patterns). This changeover can be delayed by extended overcast weather. Second, juvenile birds move from a reliance on information gathered during their flight (or experimental transport) to the homing site, and instead increasingly depend on data collected at the release location itself. Experience, either in voluntary excursions or active releases, accelerates this changeover in map strategies; confinement in the loft can delay or eliminate it. Age and experience can be equally important in migrants. Young birds learn about celestial cues, create a local-area map, and (either before or during the first migration) collect enough data at least to act as though they have large-scale spherical maps.

Species matters too. Swans, ducks, and a number of other waterbirds, for instance, lead their young south that first autumn and the juveniles learn the route as a strip map; they fly in groups, and only rarely at night. The residential populations of Canada geese so familiar to those of us in the northeastern United States are descendants of geese that missed this critical first journey, lingering through mild winters, subsisting on the bounty of suburban fields and ponds. Unless they joined some passing migrants, the round of young reared by these homebodies had no one to show them the way south. They were thus constrained to spend the cold months in place, with short zugunruhe-motivated flights from pond to pond their only reminder of the species' normal behavior. Most long-distance migrants, on the other hand, travel individually and at night; juveniles are on their own, and thus need to know something about where they are going in advance.

One way to ask migrants what they know is to intercept them en route to their seasonal goal and then transport them somewhere else. If they depart from the new site along their original heading, they are performing vector navigation and need not have a map. If they head back to the capture point they may be behaving like homing pigeons, returning to the starting point and then presumably taking up their journey again along the usual migratory corridor. This would require some sort of map or an inertial ability to measure their displacement. But if the kidnapped birds fly directly to their ultimate goal, adopting a novel route to get there, they must have a true map, an effective wide-area GPS of some sort.

To exclude the possibility that juveniles may be using different cues and navigational strategies, the obvious trick is to look at spring migration. All birds flying north in spring are veterans of one or more journeys between summer and winter ranges. When this was tried with reed warblers migrating NE into the Russian Arctic, most birds transported 600 miles east attempted to leave their new location heading NW toward their original target, demonstrating a true and large-scale map sense.

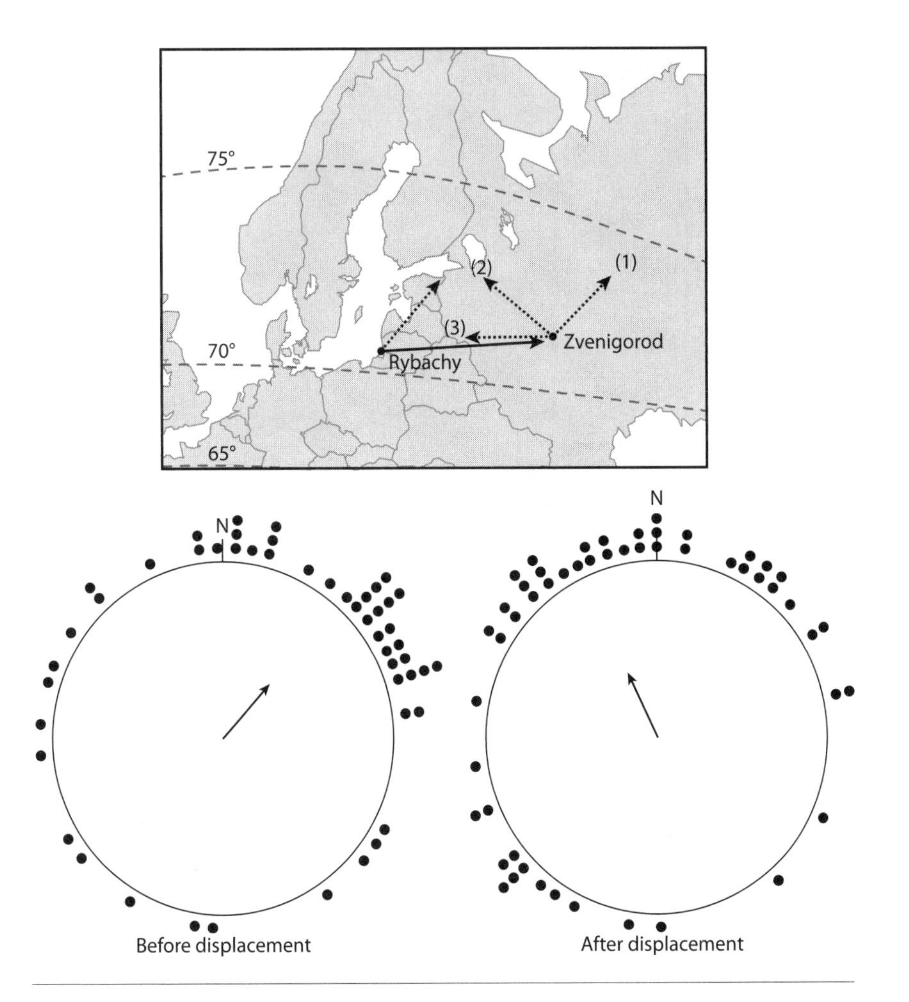

Displacement of adult migrants. Reed warblers en route to their Arctic breeding grounds were captured along the Baltic Coast. Tested in cages at this site, they attempted to continue their trip to the NE. The birds were then transported in sensory isolation by plane 600 miles to the east and tested again. If they were performing vector navigation they would be expected to continue flying NE (1). To return to the capture site and resume their journey, they would need to head west (3). Instead, the warblers attempted to fly NW toward their breeding grounds (2).

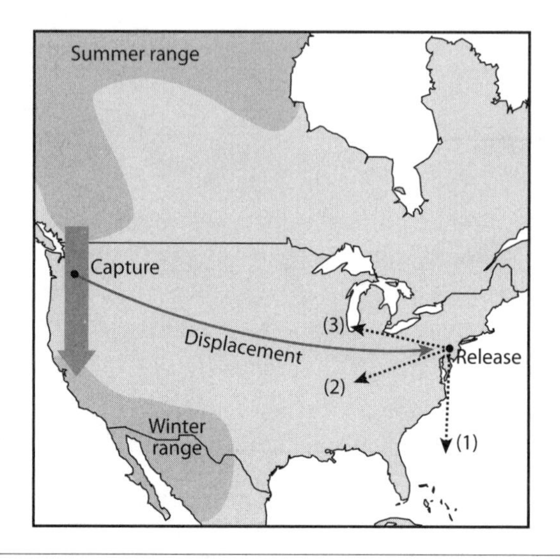

Transcontinental displacement test. White-crowned sparrows captured in the western United States were flown in darkness more than 2200 miles to the ESE for release. Three alternative predictions are shown: (1) continuing south (vector navigation), (2) flying directly to the goal (large-scale map), and (3) returning to the capture site to continue the original route.

An analogous test in the United States enlarged on this experiment and filled in several important details. In the autumn researchers captured white-crowned sparrows in Washington State en route from their Canadian breeding grounds to the winter range in Mexico and the American Southwest. The birds were flown to New Jersey, a distance almost three times that of the Russian tests. Juveniles, recognizable in this species from their immature plumage, were considered separately from the adults, allowing an analysis of the role experience plays. Radio tracking removed any worries about the artificiality of experiments with caged birds.

Again, the transported birds could do one of at least three things. They could continue vector navigation by flying south; travel WNW back to the capture site in order to continue their route-based strategy; or set off WSW to the actual wintering grounds. Both adults and juveniles flew around locally for a time, perhaps feeding as they would for a migratory stopover or just recovering from the stress of their kidnapping. Soon, however, they began to depart the area. Juveniles flew south with a 99% consistency in their bearings, a clear indication that they were using vector navigation. Nothing in their behavior suggests that they realized that they were on the other side of the continent; perhaps they did not understand that they had been displaced or, more likely, they did not know what to do about it.

Adults responded in a completely different way. Except for one that was blown to the shore by a strong 25-mph wind from the NW (the other adults wisely chose not to fly until the breeze abated), the experienced sparrows set off bearing strongly WSW toward their wintering grounds some 2000 miles away. Their map ability evidently extends over a huge range. Other tests indicate that migrants can be moved at least 6000 miles and still orient to the target. The shift from a vector strategy—flying a rhumb line—to map-

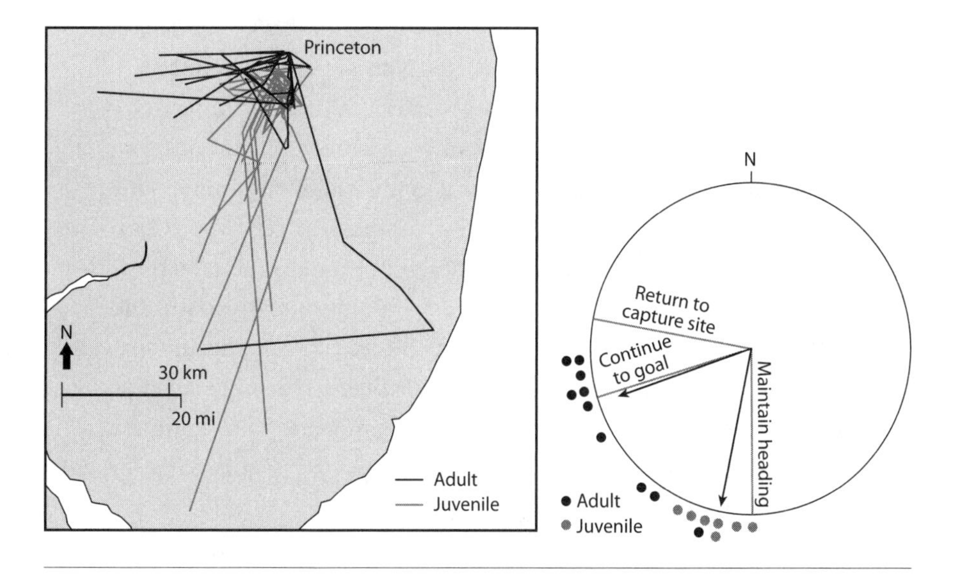

Behavior of displaced sparrows. After spending time in the vicinity, juvenile sparrows departed south along the vector they had been flying before transport. Adults, by contrast, flew to the WSW toward the actual goal.

based navigation (when birds switch to great-circle routes) is typical of many migrant songbirds.

Any serious large-scale–map theories need to account for this essentially unlimited global range, the remarkable consensus among animals about the direction from the current location to the intended target, and the capacity of the genome in certain species to encode enough information about the goal to permit a well-calibrated organism to know innately when it is getting close. Accounting for the vector-to-map shift—the need for experience to enable the full-fledged map strategy—is another responsibility of any viable hypothesis.

■ True Maps

There are only two general types of "true" maps that navigation researchers consider capable of explaining homing and pinpoint navigation. The first is the kind we are used to. Let's suppose, in the best traditions of Hollywood, that you are kidnapped, blindfolded, and taken to an unfamiliar location to be held for ransom. Then you escape. The conventional view is that to get back you will need a map to tell you where you are relative to home—the distance and direction—and then a compass to set your return course based on this map information. For humans the map position is defined by a latitude and a longitude: north–south and east–west measurements, respectively. These can be as accurate as the sensory equipment permits. But there is no reason to suppose the parameters used by homers and navigators are in fact aligned along geographic axes, or that they intersect perpendicularly, or that there are just two.

Another possibility is that there are instead radial cues, so that a pigeon at its loft senses some unique cue (or blend of cues) from each compass direction. Displaced from the colony, each bird

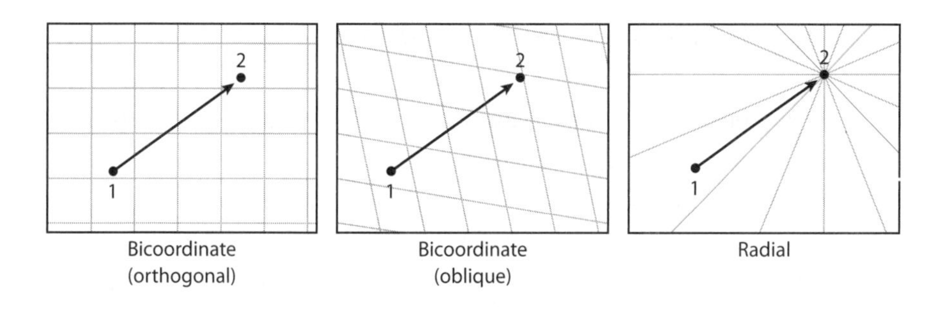

Bicoordinate (orthogonal) Bicoordinate (oblique) Radial

Alternative map systems. Here are the two proposed map hypotheses that might account for how animals return from large-scale displacements or accurately locate distant targets. A bicoordinate system requires the position of both the starting point and the goal—the latitude and longitude in the version at the left. From these two points the net distance and direction can be inferred. A compass is then needed to orient. But the two coordinates need not vary at right angles to one another, as the oblique version in the center indicates. The main alternative is a radial map, which requires unique stimuli arriving at the goal from each direction; the stimulus experienced at the displacement site informs an animal as to its direction (but not necessarily its distance) from the target. Again, a compass is then needed to set course. Presumably the animal depends on encountering local cues to recognize when it reaches the goal, or (presuming the intensity of the radial cue declines with distance) it might use the strength of the orientation signal as a guide to distance.

would determine its direction by monitoring this cue, and then take up the appropriate return bearing by means of a compass. The radial map does not necessarily provide distance; an animal might need its local-area map to recognize when it is near home. Exactly how the transported white-crowned sparrows could use such a system is not clear; some cue on the East Coast would have to correspond to a signal emanating from the American Southwest. The adults would need to have mapped these radial cues during the previous winter and extrapolated them 2000 miles in each direction.

■ The Longitude Problem

We map our spherical planet using a grid of latitude and longitude lines, divided into degrees. There are 360° of longitude along any east–west latitude line. If you were to circle the globe north–south beginning at the north pole, continuing to the south pole, and then returning on the opposite side of the globe you would also cross 360° of latitude. Each degree is divided into 60 minutes; each minute of arc is subdivided into 60 seconds. By convention latitudes are measured north or south of the equator (latitude 0°); longitudes are measured east or west of a prime meridian running north-south through the Royal Observatory in Greenwich, Great Britain. Thus, for instance, the *San Antonio* went aground at 32°15′17″ north latitude, 64°54′15″ west longitude.

Determining latitude is trivial: the elevation of the pole point, for example, gives it directly. Working out longitude is far more

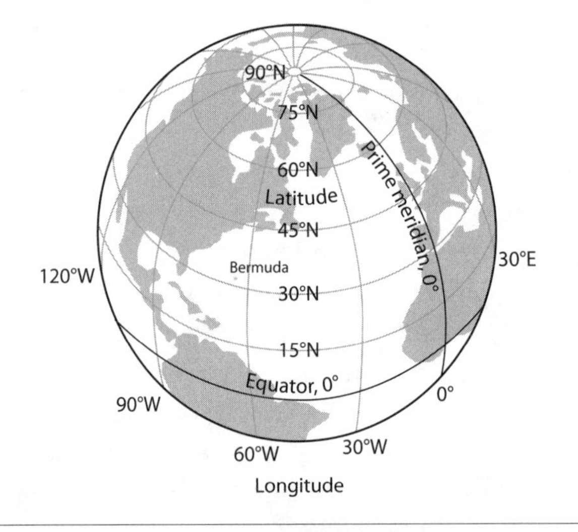

Latitude and longitude lines. Human navigators map the globe in terms of degrees of longitude and latitude, measuring latitude from the equator and longitude from the prime meridian. Bermuda is shown at approximately latitude 32°N, longitude 65°W.

challenging. The only difference between a patch of ocean at 35° north latitude in the Atlantic versus 35° north latitude in the Pacific is the time of day—a difference of 3.5–11.5 hours depending on the two locations being compared. And the only way a sailor could measure this value before the United States began broadcasting worldwide time signals was with a highly accurate timepiece—a practical impossibility before 1800.

In analyzing the map abilities of animals, humans naturally assumed that long-distance homers and migrators would use time to judge relative longitude. The idea is simple: a creature could compare the local time of day (based on the sun's position, or the corresponding pattern of polarized light, or the stars at night) with an internal clock that provided the reference time at home. Although as we saw in an earlier chapter, internal time (as measured by other behaviors) is not very precise, drifting about a minute per hour at best, surely this would be good enough to allow homing pigeons to deduce their location after a couple of hours of travel in sensory isolation. (An error of one minute per hour corresponds to 15 nautical miles of longitude; for comparison, we saw that the scatter in release bearings after a 50-mile displacement is equivalent to an initial ambiguity of about 25 miles.)

A single test is sufficient to show that birds do nothing of the kind. In a classic experiment, Schmidt-Koenig simply clock-shifted pigeons six hours fast so that their internal clocks read 6 p.m. at solar noon. When transported south in the dark and released at noon, they had a choice. If they navigate like human sailors in the 1800s they would observe the sun high in the southern sky. There is only one place on earth that the sun is at that elevation when the time at home is 6 p.m.: 90° to the west, in the Pacific north of Hawaii. This being so, the bird should fly east. But if the bird knows its true location in some independent way, the sun is merely a compass to be used to orient back north to the loft. As a celestial landmark that should be in the west at 6 p.m., the sun tells the pigeon to fly 90° to the right (CW) of it to steer north. Because

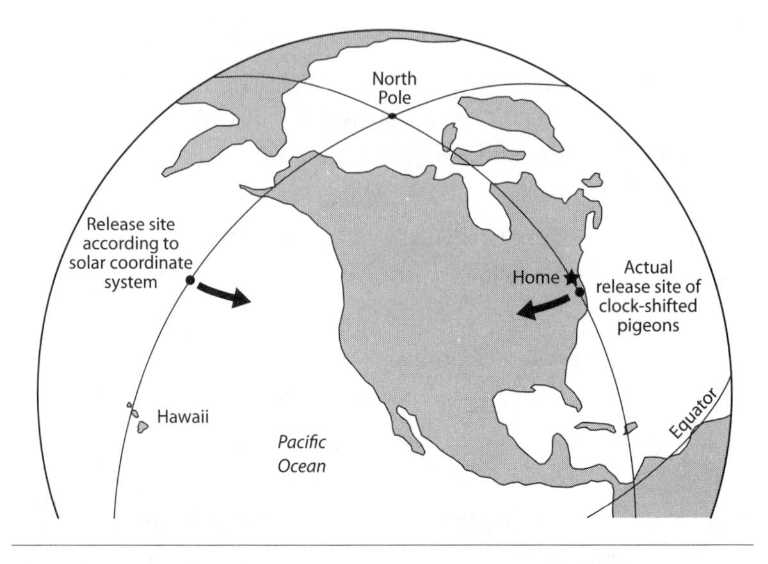

Longitude and clock-shift tests. Experiments show clearly that animals do not use time to judge their east–west position, but instead employ their clocks to calibrate celestial compasses. In this test pigeons clock-shifted six hours fast are taken south and set loose at noon. If they used the sun as a map they would infer that they were in the Pacific and depart east. Instead they know their location in some other way and use the sun as a compass: they judge the southern sun to be in the west, and steer 90° to the right (apparent north), which in fact takes them west.

the sun is actually in the south, this means the pigeons will depart to the west. The two models lead to predictions 180° apart, and both 90° from the true direction. The results are unambiguous: birds—homing pigeons, bank swallows, and every other species tested to date—set off to the west. Animals do not determine longitude from clocks.

■ An Olfactory Map?

If the map is not celestial and is independent of time, what can animals be doing? Given our species' harrowing trials trying to figure out east–west position, it's easy to see the attraction of the longitude-free, radial-map idea. The evidence for such a strategy began with a set of anomalous observations—data that might not mean anything, or might instead hold the key to the mystery. These puzzling discrepancies grew out of the work of Gustav Kramer, the first researcher to show that birds have a sun compass. Kramer was investigating the ontogeny of the compass—how birds come to "understand" the sun's arc though the sky. To control what his homing pigeons could see of the heavens he constructed so-called palisade lofts, enclosures surrounded by walls that blocked the young birds' sight lines.

As we saw in a previous chapter, this rearing technique denies the juveniles any flight experience, which unfortunately interferes with their ability to imprint on the loft. This deprivation also blocks or delays the maturation of the sun compass, and presumably denies them the experience that in normal pigeons leads to the ability to deduce location at a release site rather than from cues en route from home. To his complete surprise, Kramer discovered that if he blocked the young birds' view of the horizon and the 3.5° above it they were very poorly oriented at release sites. *First-flight* control birds reared in lofts with the horizon visible were less scattered, displaying at least a weak preference for the homeward direction; *first-test* pigeons raised under natural conditions are more accurate (plus they mostly get back to the loft after this initial release). Intrigued, Kramer tried a variety of alternative lofts. A partial view of the northern horizon did not help much. A gap showing a small part of the southern horizon was not much better. A palisade with a view of the horizon through glass windows yielded intermediate performance.

If instead the palisade was roofed but allowed a direct, clear view of the horizon and the few degrees above it, the birds were no worse than when they were imprisoned in a fully open loft. This was true even if the horizon was visible only in parts—that is, a different segment each day. These results were very perplexing at the time, particularly the fact that glass seemed to be about half as effective as plywood in blocking (or reorienting) some essential cue. In 1966 Kramer's student Hans Wallraff summarized the results thus: "barriers between the aviary and the external world interrupt or change a substrate of information spreading in the horizontal plane which is important for homing ability. Its nature is unknown."

Today we understand the potential importance of sunrise and sunset, both ultraviolet and polarized light, and in particular the

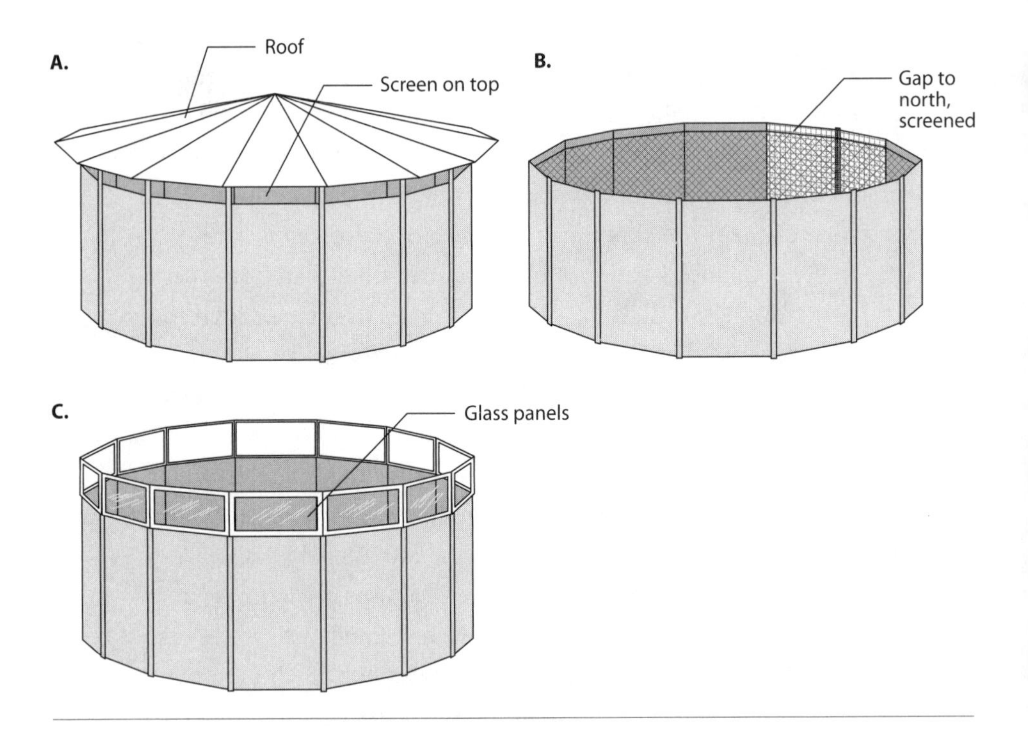

Palisade lofts. The roofed palisade (A) allows the birds a view of the horizon from the top-most perches. The gap palisade (B) permits the juvenile pigeons a sight of the northern horizon plus most of the sky, but not the eastern, southern, or western horizon. The glass palisade (C) provides a view of the horizon, but only through glass. The birds in the roofed palisade show a modest ability to orient at release sites, the glass-loft birds less, whereas the gap-loft birds are lost.

effects of glass (both its transmission and reflection) on polarization and ultraviolet light. Given the normal ontogeny of pigeon compasses and maps, numerous alternatives spring instantly to mind. At the time, however, the results presented a daunting but intriguing, deeply seductive mystery.

In 1972, the Italian researcher Floriano Papi and his colleagues suggested that this "horizontal factor" might be odor. Birds in the loft, he hypothesized, might memorize the prevailing odors arriving from each compass direction. Then when taken from the loft they would need merely to sniff the air at the release site to determine the direction in which they had been transported. Using the sun as a compass, they would set off in the homeward direction—presumably with no knowledge of distance. Once near the loft an olfactory homer would count on using its local area map to complete the journey. Given that pigeons without form vision can get to within a mile of their loft, perhaps this short-range locale information could be olfactory as well. Alternatively, because the intensity of odors doubtless declines with distance through simple diffusion, there might be distance information encoded in this olfactory signal after all.

The experimental evidence he offered was dramatic: surgical severing of both branches of the olfactory nerve in the beak reduced initial orientation of first-flight birds to the level of palisade pigeons; cutting just one side had less effect. Papi could not have known that another main contender for the pigeon map sense sends its information through an adjacent nerve.

The idea that animals might have odor-based maps was very much alive at the time. In fact, the whole incredible story of the honey bee dance language had been vigorously challenged since 1967. New, seemingly better-controlled experiments appeared to show that recruits found the food advertised in the dances on the basis of odors absorbed onto the waxy hairs of the dancers. The angle of the waggle and its duration were taken to be interesting but useless correlations, no more meaningful than the well-known numerical connection between temperature and the rate at which male crickets chirp. In 1975, however, we were finally able to separate the dance coordinates from the olfactory information. Recruits, it turns out, can use either. They choose a familiar odor when it is strong and unambiguous, but use the dance information under most normal circumstances. If bees can employ odor maps under certain conditions, why not birds?

In an ingenious series of experiments, Papi and his colleagues built so-called deflector lofts. These structures had large glass or Plexiglas vanes designed to rotate the incoming wind. If airborne odors are key, this should rotate the olfactory map as well; in consequence, the birds should show a corresponding reorientation at release sites. And, indeed, this is just what they found—at least on sunny days. Later, however, other researchers showed that most or all of the effect was created by the reflection of the sun and sky off the vanes.

Other seemingly conclusive tests involve filtering the air over activated charcoal to remove organic scents. (The air can be scrubbed in real time, or bottled earlier and then metered into the chamber on the way out; it doesn't seem to matter.) For these tests the results seem clear-cut: first-flight pigeons given odorless air on the way to release are poorly oriented at longer distances (more than 50–100 miles). Control birds, having breathed ordinary air, home normally. But recently John Phillips and his colleagues discovered that the problem is probably one of motivation or activation: if arbitrarily chosen artificial scents (lavender, camellia, eucalyptus, rose, and jasmine, irrelevant to the actual journey) are added in sequence to the filtered air on the way out, the pigeons depart normally. It's not location-specific odors that seem to matter to them; the birds just need something in what they breathe that the filtering removes—an olfactory wake-up call of some sort.

Impressive as some of the olfactory evidence may seem, there are some other facts to keep in mind. First, at the sensory level, birds in general and pigeons in particular have relatively impover-

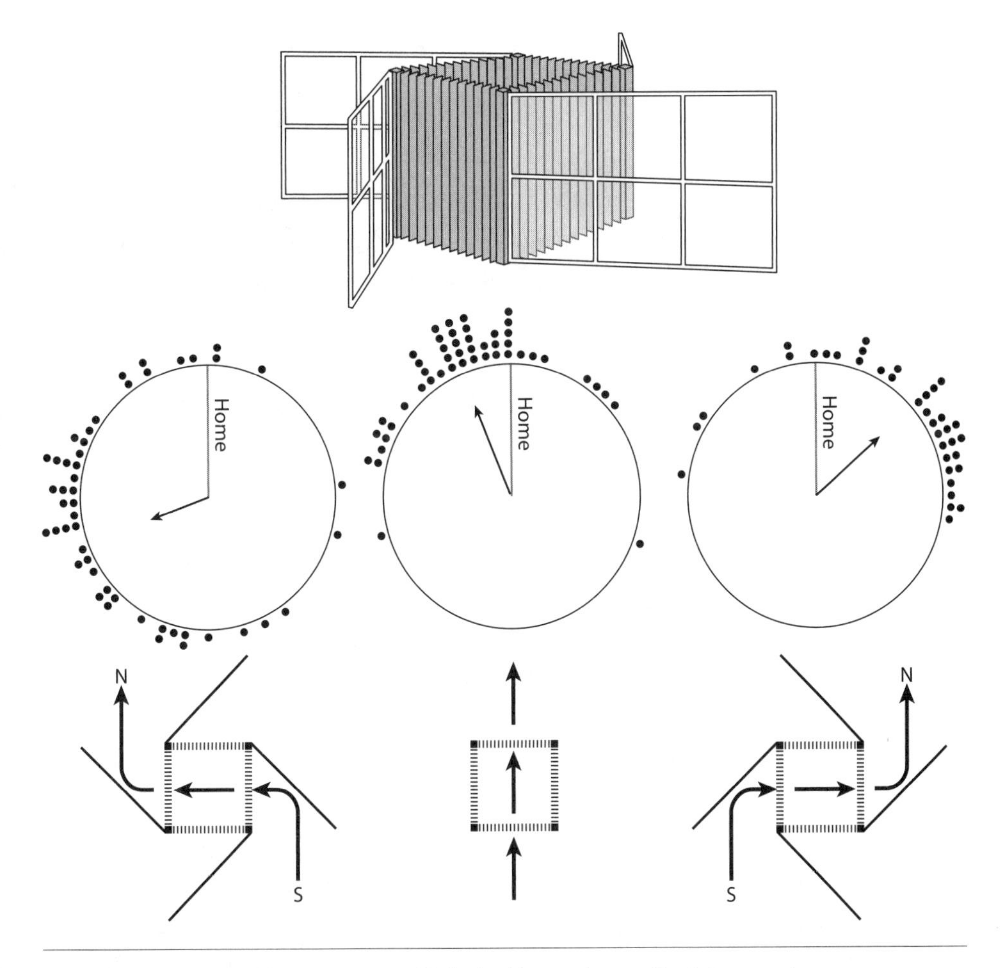

Deflector lofts. The experimental lofts had large glass or Plexiglas panels to rotate the wind CW or CCW. When the first-flight birds were released a clear effect of the treatment was evident.

ished olfactory epithelia and corresponding brain areas. Attempts to condition pigeons to naturalistic odors in the lab have failed. Second, birds reared with wind from only a single direction are nevertheless able to home from any orientation—that is, from bearings whose odors they have never sensed. Studies mapping the path of air arriving at lofts from a spot at a particular bearing

200 miles away show a low directional correlation; moreover, the errors predicted in departure angles based on odor greatly exceed those actually observed; assuming the birds are not cheating, they cannot be using scents alone.

Another approach has been to bait the air arriving at the loft with strong artificial odors—turpentine, olive oil, or benzaldehyde. The wind from one direction is laced with one of these experimental scents, and a different odor added to the air from another quarter. When taken to a release site and exposed to the odor en route, or with the odor painted on its beak before being tossed, a pigeon is likely to fly back in the direction it associates with the smell. Here, finally, is an effect that does not go away under overcast, requires no surgery, does not mysteriously appear only at great distances, and does not require an elaborate averaging of averages to wring statistical significance from the data. Quite simply, birds reared without flight experience can be conditioned to associate strong odors with specific directions. Whether this is relevant to normally reared birds in a world of natural scents, however, is another question.

But the most worrying problem is one of scale. When we think back to the transcontinental displacement of migrants, we must ask how odors on the East Coast of the United States could provide any information about the direction of a wintering ground 2000 miles away to the WSW. Even if the white-crowned sparrows in that test knew the odors of northwestern Mexico from their experience on the West Coast, what could this possibly mean to them after transport to—for all the bird knows—Hawaii to the west, or Canada to the north, or Panama to the south? The birds have never had an opportunity to associate odors with directions in New Jersey, and yet their departure bearings were tightly clustered. Surely other cues are being consulted. Nonetheless, interfering with the olfactory systems of pigeons and other avian migrants often does affect the quality of their orientation in some way.

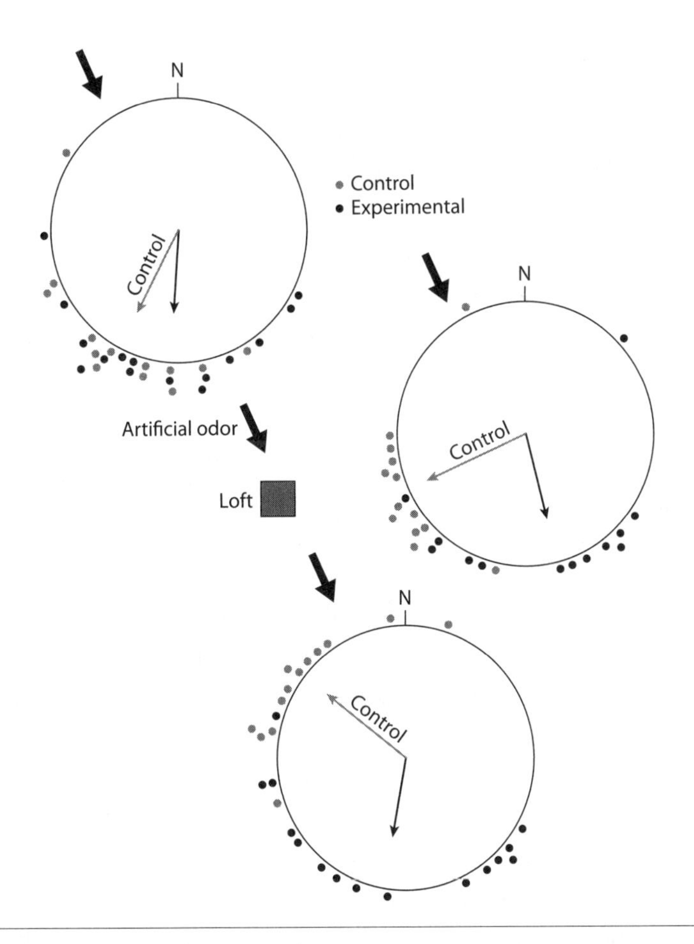

Baited-air tests. The birds were reared with winds from the NNW strongly baited with benzaldehyde, with fans used at times to increase the strength and consistency of the exposure. First-flight pigeons were released roughly 30 miles away in three directions. Experimentals (whose departure bearings are shown as dark circles) were exposed to the training odor en route, at the release site, and even while flying home; they tended to fly south regardless of location. Controls (light circles) were generally homeward oriented.

■ A Magnetic Map?

By 1980 the first of many shortcomings in the olfactory hypothesis had become clear. Independently we, along with Bruce Moore, a psychologist at Dalhousie University, and Charles Walcott, were drawn to a different set of perplexing anomalies evident in the orientation behavior of conventionally reared pigeons. We each proposed that the map is magnetic, and developed fairly similar hypotheses to account for these apparent oddities in bird orientation behavior.

The unexpected effect of solar storms on homing ability was the first curiosity in the data to suggest a magnetic influence. As we mentioned in an earlier chapter, the large and relatively static magnetic field produced inside the earth is supplemented by the movement of charged ions through the atmosphere, especially in the jet streams. This small induced field has a fairly regular daily pattern that is interrupted by magnetic "storms" resulting from solar flares. About eight days after a solar eruption enormous numbers of new ions join the jet stream, distorting the normal pattern (and also producing the northern lights). Typical daily variability in total field strength is about 20–30 nT (nanoteslas, also called "gammas"—γ); storms increase this value up to a hundredfold. (The record was a 1600 γ storm in 1859.) Even so, this is a small value: the static background field generated by the core of the earth is on the order of 50,000 γ. But given that honey bees can use this cycle to reset their internal clocks, an animal's theoretical ability to detect at least a 5 γ change if selection has been at work is not really at issue.

Pigeon racers had long claimed that their birds homed more slowly and less successfully on days of strong magnetic storms. Could this anecdotal effect be real? Researchers went back to their release data and reanalyzed it. As you may recall, each location has a characteristic release-site bias; experienced pigeons depart reli-

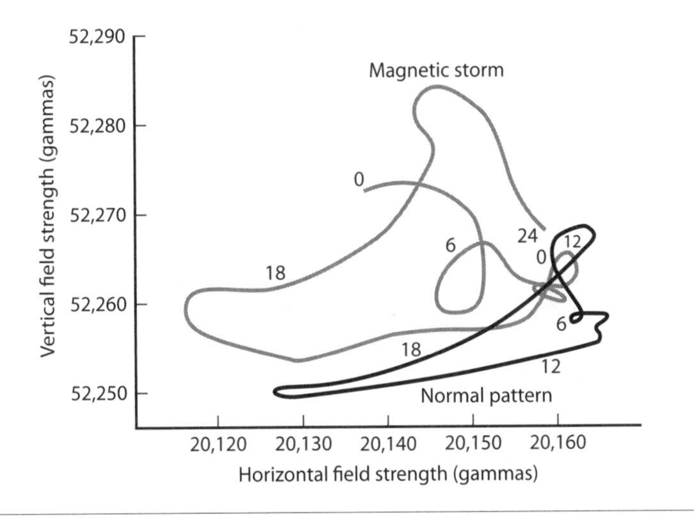

Daily magnetic variation. A typical daily pattern is shown at the lower right, along with the variation induced by a very weak storm. (Time of day is shown in 24-hour format.)

ably a bit (often quite a lot) CW or CCW of the homeward bearing. The exact degree of bias varies from day to day, but there is a consistent twist associated with each site. Tracking the birds farther from the release site shows that the bias decreases with distance. Attempts to use clock shifts to "aim" homers in the correct direction lead to lost birds and long return times: the bias must make sense to the pigeons.

It turns out that the bias on a particular day is well correlated with the ongoing magnetic activity overhead: a strong magnetic storm can rotate a pigeon's departure by an additional 40° or more. This is not an effect on the birds' compasses; these are sunny-day tests. In addition, the change in horizontal field strength needed to generate a 40° rotation is at least 12,000 γ, ten times the actual value of even a strong storm's added field. With their compass clearly visible in the sky, it's as though the pigeons are misreading their map location.

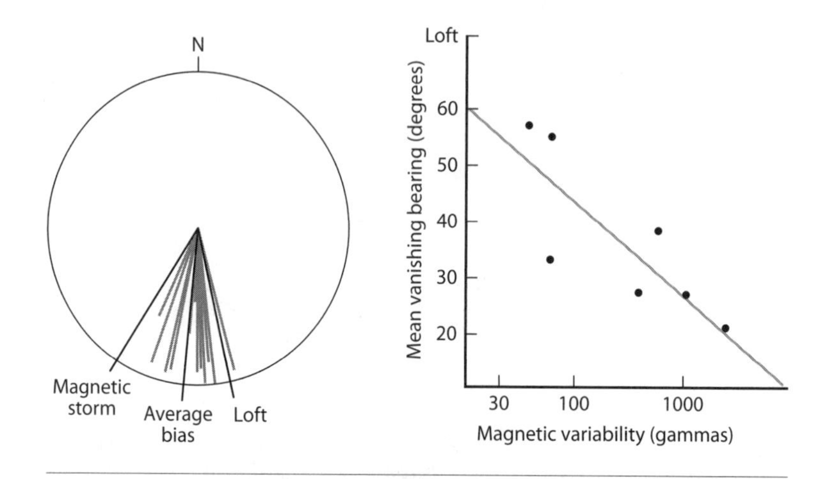

Release-site bias. (Left) The site shown here has a 15–20° CW bias on most releases, though there is variation from day to day. (Right) When the variations are plotted against magnetic variability the effect of this small parameter becomes clear.

When we look more closely at the release-site biases around well-studied lofts, another clear and exciting pattern emerges. Consider Bill Keeton's pigeon loft near Ithaca, New York, for instance: the biases to the right of a SSE–NNW line passing though the loft are typically CW; those to the left are usually CCW. It's as though, on average, departing pigeons believe their home is NNW of its true location. The axis of increasing magnetic field strength in this region also runs NNW. In northern Europe, birds from Frankfurt act as though their home is NNE of the loft, along the NNE–SSW axis of release-site biases. In Bowling Green, Ohio, the axis is NNW–SSE. (Cage-reared birds display idiosyncratic preferred compass directions rather than regional biases, an indication that flight experience is necessary for release-site biases to develop.)

An even more compelling piece of evidence at the time was literally an anomaly. Walcott had recently tried releasing pigeons on a clear day at a magnetically disturbed location atop an old iron mine. Though the sun was available and the strength of the anomaly was sufficient to rotate a compass needle only a few degrees in any event, the birds were very poorly oriented. Once well clear of the anomaly, however, the individuals recovered their bearings and turned toward home. Releases at magnetically normal sites

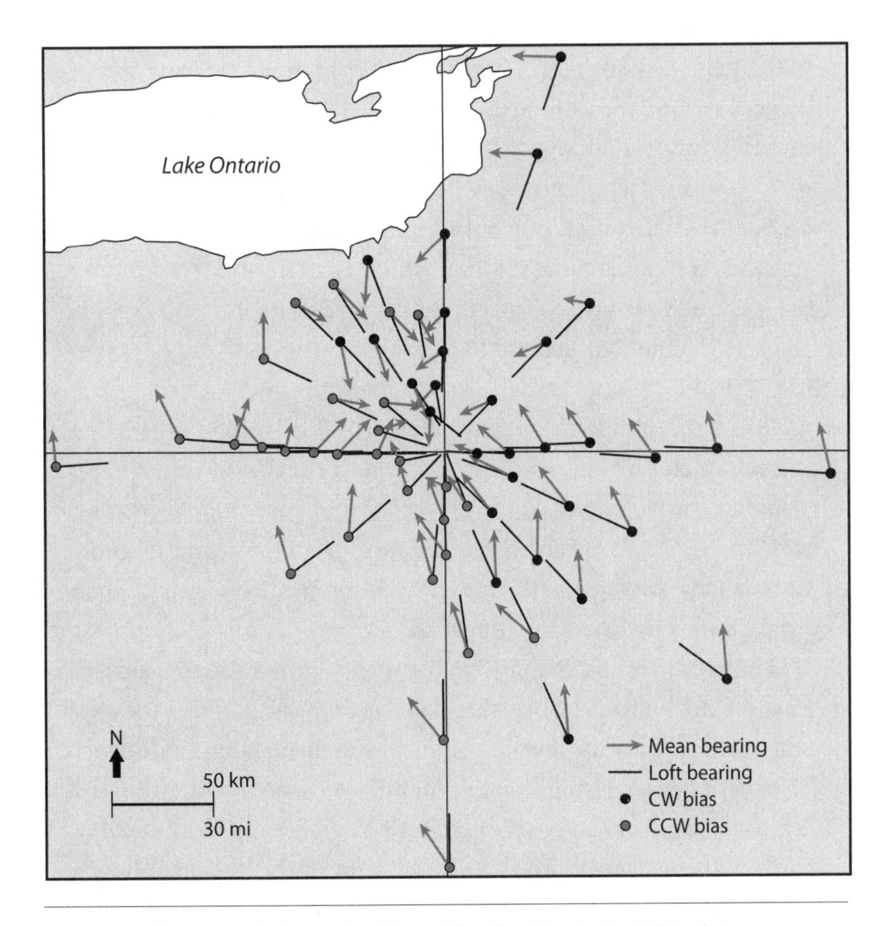

Regional pattern of release-site biases. The direction to the loft is shown as the longer bar radiating from each location. The mean departure direction is the shorter line, with the length corresponding to the length of the mean bearing. CW biases are indicated in the darker gray. Note that some of these consistent biases are as great as 60–120°. Those to the right of a line passing SSE–NNW through Ithaca are generally CW; those to the left are typically CCW.

yield much better initial orientation. As Walcott went on to show, the effect is "dosage dependent": more-irregular anomalies generate more initial confusion in the birds. The most obvious interpretation of these tests is that the location sense of homers is greatly disturbed by magnetic-field anomalies.

The time-independent map, then, seems to be partially or entirely magnetic. The problem is that all of the obvious components of the field vary (roughly) north–south, providing (at least at first glance) redundant estimates of latitude. But a closer look reveals that this is not quite true. One potential component is *total magnetic intensity*, which increases from 30,000 γ at the equator to 60,000 γ at the poles, but not exactly along a north–south axis. Another is magnetic inclination, which rises from 0° to 90° over the same 6000 miles, but at a distinct angle to the intensity gradient. The changes correspond to 5 γ and 0.9 minutes of arc per mile, respectively. Yet another is *vertical intensity*, which intersects the total intensity gradient in most places at about a 30° angle; using this parameter would require measuring vertical with some precision. The last obvious candidate is the *intensity slope* direction, which is generally oriented in the range of 60–90° from latitude. Clearly any bicoordinate map involving magnetic parameters would have a distinctly oblique grid.

How could an animal use such a map? For homing pigeons, at least, it's not enough to measure the static values at home; the animal needs to know how values change with distance and direction. Thus experience during exercise flights would be essential. Learning that, say, total intensity increases at 5.5 γ/mile to the NNW while vertical intensity rises at 10 γ/mile to the NW would allow a creature to extrapolate into the wider world before it gains direct personal experience at greater distances. Release-site biases would be compounded of several components—in particular any discrepancy between the local, loft-centered gradients and the area-wide pattern, plus any deviation between values at the release site and those predicted from extrapolation. The first potential dis-

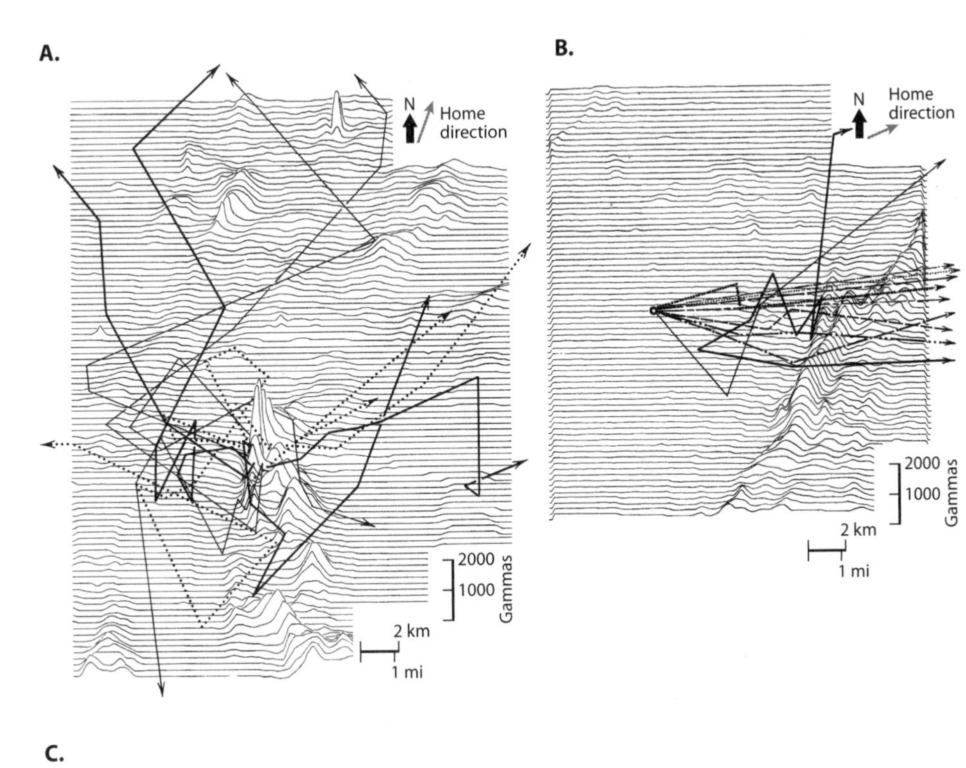

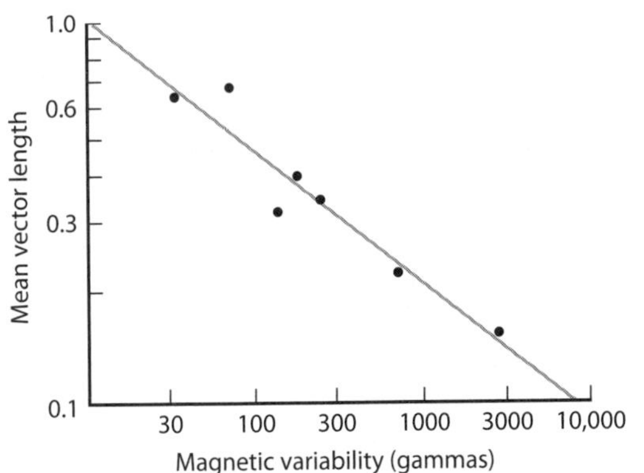

Effects of magnetic topography. Total magnetic field strength is represented here as contour lines. The normal pattern in the northeastern United States is a gradual increase in strength to the NNW. (A) At the iron mine the pigeons are initially disoriented. (B) At Worcester, Massachusetts, where the field is relatively normal, departure bearings are far more accurate and consistent. (C) The more extreme the anomaly, the more scattered are the departure bearings (leading to shorter mean-vector lengths).

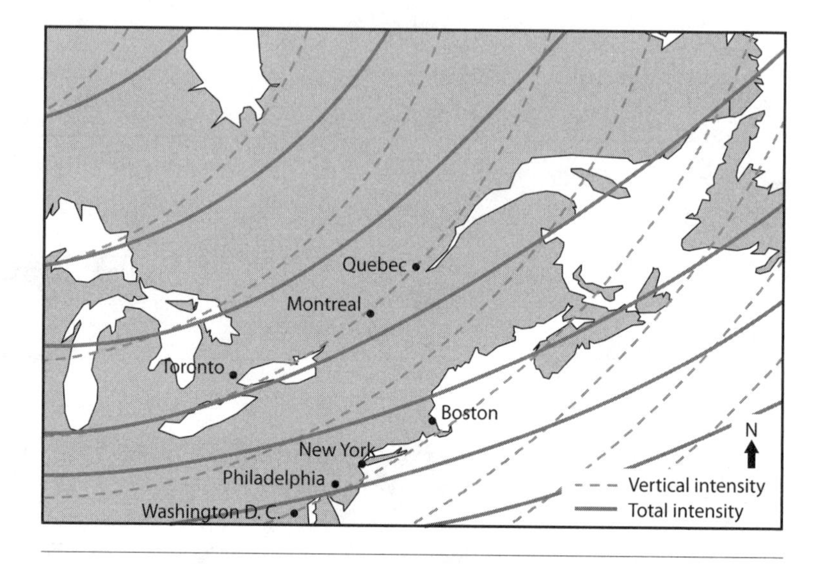

Magnetic gradients. The gradients of total and vertical magnetic-field intensity intersect at about a 30° angle in the northeastern United States.

crepancy would provide the axis and CW–CCW rotations of the general bias pattern; the second would supply the site-specific degree of rotation (the unique component of the otherwise puzzling release-site bias).

If animals have magnetic maps, how would they sense the small differences in field strength and direction necessary for this strategy? The putative detector for magnetic information in vertebrates is the magnetite-rich organ in the ethmoid sinus of the nose or beak. It is innervated by the trigeminal nerve—the one that runs adjacent to the olfactory nerve, an accident of anatomy that caused problems when researchers sought to block one but not the other. Recordings have detected the output from the magnetically sensitive organ in many species; in at least mole rats and European robins the projection of this organ to the brain has been discovered as well. Histological studies have revealed the chains of magnetite crystals that form the structure of this organ. Careful work by some German researchers using localized magnets, anesthetic, and surgery on homing pigeons confirms that this organ is critical to the map step of homing and navigation.

One wild card is the recent discovery of another putative iron-based receptor in the head. Japanese researchers have identified magnetite in the inner ear of pigeons—an organ involved in balance, orientation, inertial navigation, and hearing. Vertebrates have dense crystals of calcium carbonate called otoliths in special inner-ear chambers, the saccule and utricle. The otoliths rest on a bed of sensory hairs where their deflection of these structures allows the nervous system to compute the direction of gravity and (along with the semicircular canals) measure—and compensate for—acceleration and other head or body movements. This is how we keep our balance while walking. Birds and reptiles have a third chamber, the *lagena*, with its own otoliths. The crystals in the lagena (but not those in the saccule and utricle) have ferromagnetic domains incorporated into the calcium carbonate matrix. Recent tests by an American group show clear responses to changing magnetic inclination, both in the organ itself and downstream in the brain. Perhaps the lagena (with its ability to measure vertical movements and also compensate for lateral movements) is specialized for computing either magnetic inclination or vertical intensity, whereas the beak organ focuses on determining total intensity (which is not affected by head or body movement).

How is it that pigeons can home while wearing strong magnets? We saw that magnets create difficulties for older pigeons under some conditions by affecting the cryptochrome-based compass in the eye (though this is evident only on cloudy days when celestial information is unavailable). But surely the strong static field of the magnet would make reading a magnetic map difficult too. Later tests comparing the effects of powerful static versus weak variable magnets on migrating species show that strong fields are ignored but small changeable ones create problems. If animals are measuring small differences in the presence of a strong background—gradients, in fact—then perhaps this is what we should have expected.

Researchers have recently begun to test animals that do not fly free by creating virtual magnetic journeys. Species tested to date include sea turtles, spiny lobsters, newts, and caged birds. In lab tests researchers can alter with good precision two of the most promising magnetic parameters, total intensity and inclination. One of the most elegant tests using this approach was performed by John Phillips and his colleagues, then at Indiana University. Following up on his extensive prior work, Phillips collected red-spotted newts from their home pond and took them to his lab about 26 miles to the NNE. During testing the amphibians were exposed to a magnetic field with one of five different magnetic inclinations, ranging from 2.17° steeper to 1.83° less steep; this corresponds to a range of about 120 miles north and south of the testing site. Newts subjected to more northerly inclinations oriented south, while those seemingly displaced south oriented to the north. Clearly the animals could deduce the latitudinal shift on the basis of magnetic cues alone. By including some small inclination changes in the set of tests, Phillips was able to roughly estimate the sensitivity of the animals: for this magnetic-field component, the newts' threshold must lie somewhere between 0.02° and 0.17°, or 1–11 miles. It would be wonderful to see the results of similar tests varying total intensity.

A similar and equally spectacular set of tests was devised by Kenneth Lohmann and his colleagues at the University of North Carolina. They captured spiny lobsters from sites in the Florida Keys. These large invertebrates, also known as rock lobsters or langoustes, are nocturnal, living in cavities in coral reefs during the day. At night they emerge and travel considerable distances foraging for marine invertebrates such as clams and sea urchins, as well as the detritus and carrion typical of their distant cousins, true lobsters. As dawn approaches they retreat to their home crevice.

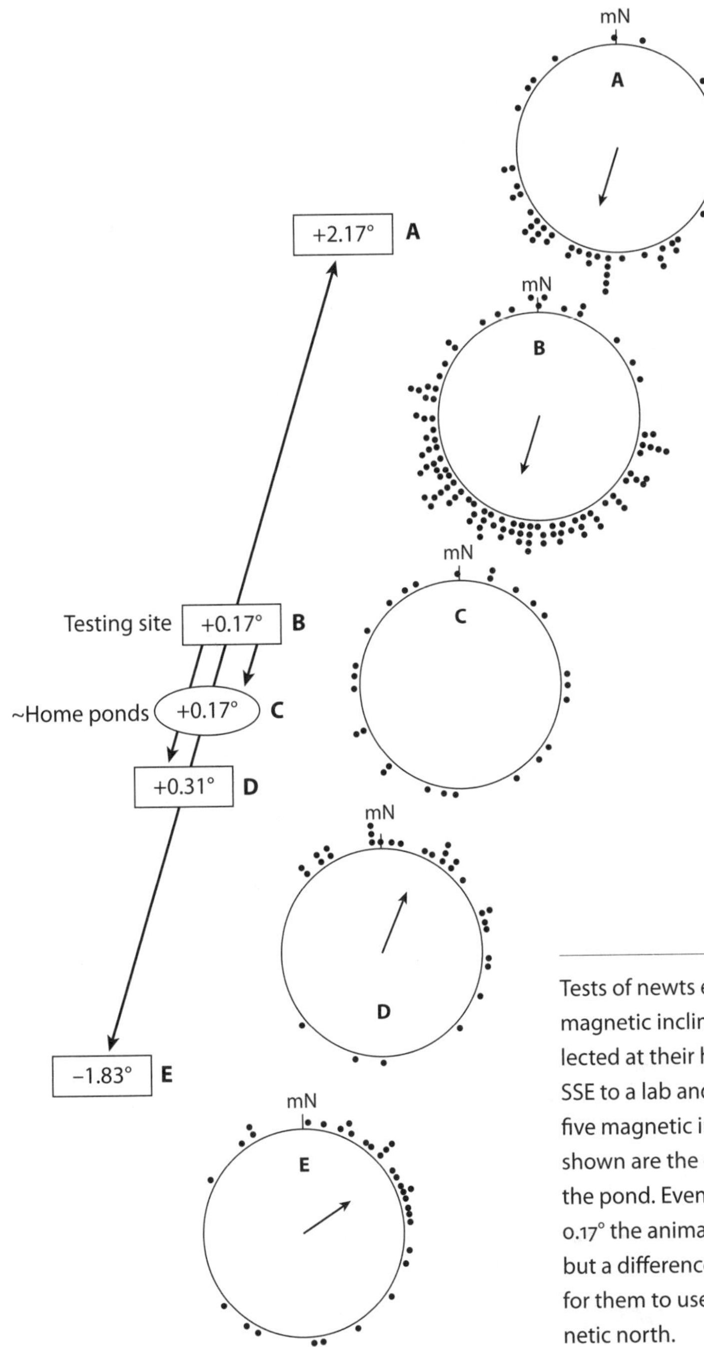

Tests of newts exposed to different magnetic inclinations. Animals collected at their home pond were taken SSE to a lab and tested with one of five magnetic inclinations. The values shown are the dip angles relative to the pond. Even with a change of only 0.17° the animals were well oriented, but a difference of 0.02° was too small for them to use. mN indicates magnetic north.

Like the newts, rock lobsters show all the characteristics of true navigation: they can home from an unfamiliar location in the absence of cues from the goal and without using information from the outward journey. For instance, taken in the dark by boat along a circuitous route, even in the presence of strong and varying magnetic fields, they can reliably choose the homeward direction from distances of 7–23 miles. And like the newts, when subjected to a virtual displacement arbitrarily dialed into the coils enclosing the testing tank, the animals correctly interpreted apparent redeployments of 250 miles north or south. (The physical displacement before testing was a mere nine miles ENE.)

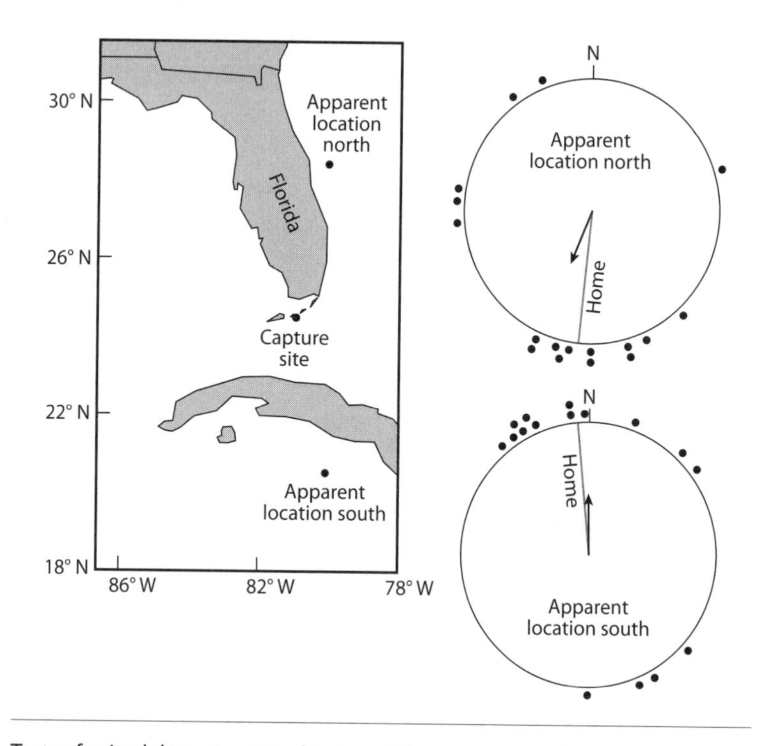

Tests of spiny lobsters exposed to two different magnetic locations. Spiny lobsters captured in the Florida Keys were taken 9 miles ENE to a lab and tested with one of two magnetic locations 250 miles distant either north or south. The animals were very well oriented back toward the capture site.

Although it's hard to see how moving the animals a few miles from the capture to the testing site could matter in this research, an ongoing series of experiments performed on loggerhead sea turtles does not, in fact, require any serious transport at all. Lohmann's group tests their animals almost where they are found as they emerge on Florida beaches. They present hatchlings with a set of magnetic parameters characteristic of a different location and ask them in which direction they would like to swim. As we will see, however, these dramatic and elegant experiments are not, strictly speaking, tests of homing ability.

Loggerheads are the largest shelled turtles in the world, weighing an average of 300 pounds as adults. They begin very modestly as 3/4-ounce hatchlings two inches long, struggling through the sand from the buried nests their mothers dug 80 days earlier. As we mentioned in a previous chapter, the baby turtles hatch at night and crawl immediately toward the brightest part of the horizon—almost inevitably the shore, where the water reflects starlight. These weak swimmers must make it well offshore before daybreak to avoid devastating predation. Loggerheads are found throughout the world in temperate and tropical latitudes, and nest all along the tropical and semitropical coasts of America, Africa, southern Europe, and islands in between. The most thoroughly studied population comes from the 70,000 nests laid annually on the east coast of Florida.

As we mentioned in chapter 5, the hatchlings from these beaches spend several years in the North Atlantic gyre, living much of the time in the mats of sargassum weed characteristic of the gyre and the Sargasso Sea ecosystem it encloses. Once the juveniles grow to about 18 inches they leave the gyre and feed along the continental shelf and in shallow coastal waters. When they reach sexual maturity at 20–30 years, the females begin producing clutches of eggs every 2–3 years. Each female comes ashore on the same beach she left as a hatchling decades earlier, excavates a pit, lays a clutch, covers it, and returns to the sea. Loggerheads can live 50–65 years.

There are several navigational problems here. The first is that the hatchling females must memorize the beach location in those first critical minutes—presumably through imprinting—so they can return to it many years later. The hatchlings must be able to maintain an appropriate and consistent direction once in the water, when their view of the horizon is blocked by every wave. They need to know when they reach the gyre and then stay within it. Finally, the larger coastal-feeding juveniles and adults must be able to return to favorite feeding spots after displacement by riptides and other currents. The initial orientation offshore involves a magnetic compass. The other three feats depend on a map sense, and clever experiments prove that only magnetic information is necessary for this ability. Because the behavior is apparent in complete darkness, it must be based on the magnetite organs that have been found in the turtles.

The ability of struggling loggerhead hatchlings to remain safely within the gyre requires them in some sense to anticipate potential problems, and take action in advance. The conditions of the test tank eliminate the possibility that the newborn animals could use currents (and animals in the water cannot sense the currents carrying them along in any event). They also rule out the option of learning and using spatial gradients of cues such as the direction and rate of change of magnetic gradients across the few miles near a pigeon's loft; the hatchlings are confined to a small tank with a uniform, gradient-free magnetic field, and have had no opportunity to learn. Thus to deal with the gyre initially, loggerheads must either be born with a preset target in the Atlantic, or they must know what to do at each specific location in the North Atlantic regardless of the latitude and longitude of the natal beach—whether they were born in Florida, Brazil, or Africa, for instance. The orientation strategy is clear: the baby turtles do not seem to have a fixed target. Instead they swim into the relative safety of the Sargasso Sea, adopting a direction that depends on their location at the moment. For hatchlings the average deviation into the Sar-

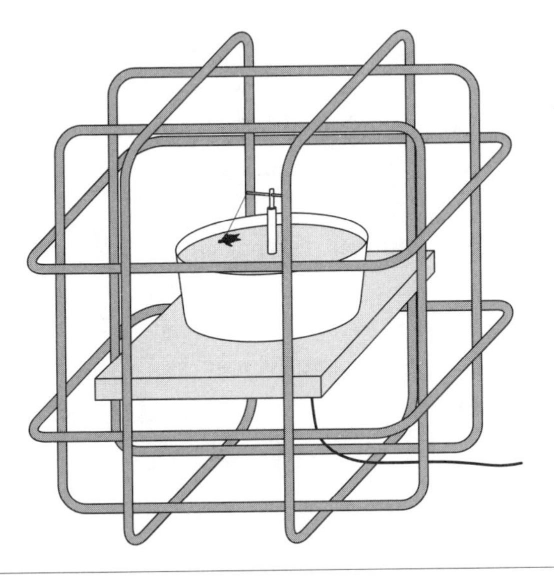

Test tank for sea turtle hatchlings. Newly hatched sea turtles are fitted with a harness at the end of a thin line, which is attached to a rotating arm in a four-foot tank surrounded by coils that create a magnetic field when energized. After 10 minutes with only a low-elevation dim light for orientation (to mimic the horizon), the light is extinguished, leaving the newborn turtles with only the magnetic information provided by the coils.

gasso is 70°—more near the parts of the gyre where dangerous centrifugal currents are found, and less along the safer eastern and western flows.

The ability of adults to return to their feeding spots after magnetic displacement also has been tested. In this case, a scaled-up version of the testing apparatus used for hatchlings was created. Loggerheads a few years old were captured off Melbourne Beach, Florida, and subjected to a change in magnetic latitude 210 miles north or south. (Specifically, the "north" group was exposed to an intensity of 49,300 nT and an inclination of 61.2°, whereas the "south" group encountered a field strength of 45,400 nT at 55.4°.) These juveniles rapidly and reliably oriented so as to return to their feeding grounds. Though sea turtles may learn about magnetic gradients during their time in the gyre, it seems clear that they can judge latitude accurately on the basis of static magnetic parameters alone. On the other hand, the longer mean vectors in these tests with juveniles may suggest that they are more certain of their location than are hatchlings.

Until recently a reasonable skeptic could remain agnostic about the reality of full-scale bicoordinate magnetic maps simply because all displacement tests either involved multiple cues (for instance, the transcontinental movement of sparrows from the western United States to the East Coast provided potential inertial cues

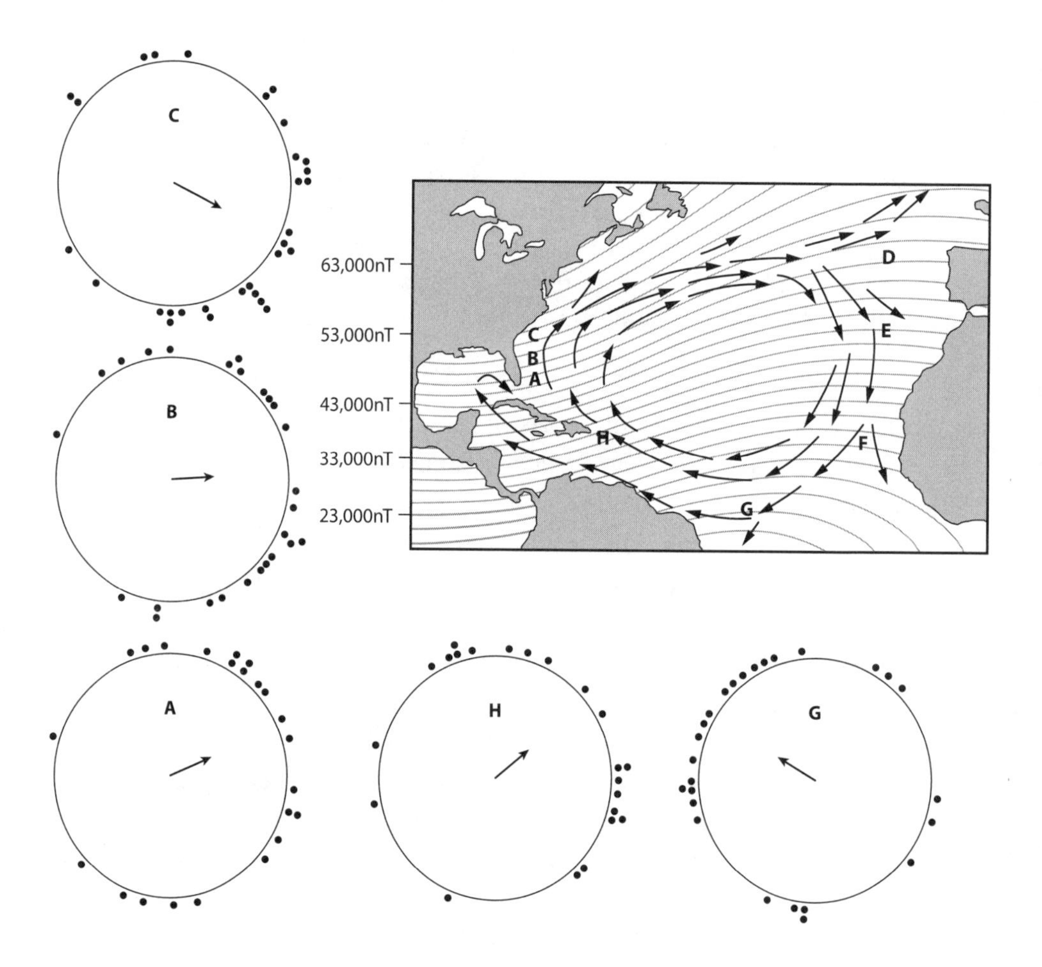

and theoretical changes in odors), or shifts in latitude that involve easily detected changes in intensity and inclination. But recent additions to the sea turtle data set seem to overcome even this worry: changing only the magnetic longitude nevertheless yields clear and adaptive orientation. In this particular test, two locations at latitude 20.0° were chosen corresponding to Puerto Rico (longitude 65.5°) and Cape Verde Island (longitude 30.5°). The magnetic parameters imposed on the two sets of baby turtles (both tested in Florida shortly after hatching, recall) were an inclination of 46.4° and strength of 39,000 nT versus a dip angle of 26.1° and an inten-

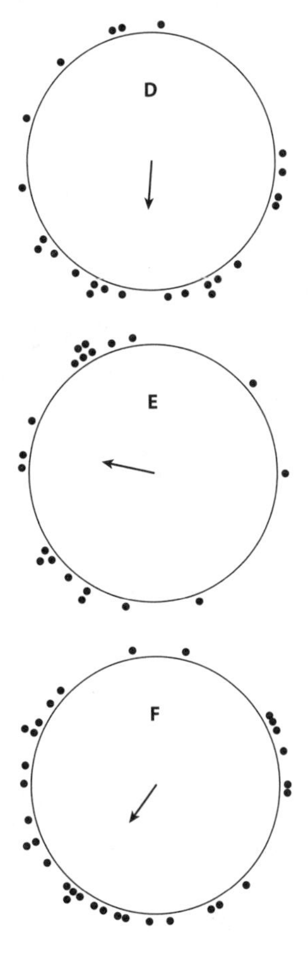

Orientation of hatchlings after virtual magnetic displacements. Newborn turtles from beaches near Boca Raton, Florida, were tested in a nearby lab, but were presented with a combination of magnetic intensity and inclination characteristic of one of the eight locations across the North Atlantic shown here. On average, the hatchlings oriented their swimming into the Sargasso Sea and away from currents that would take them outside the gyre.

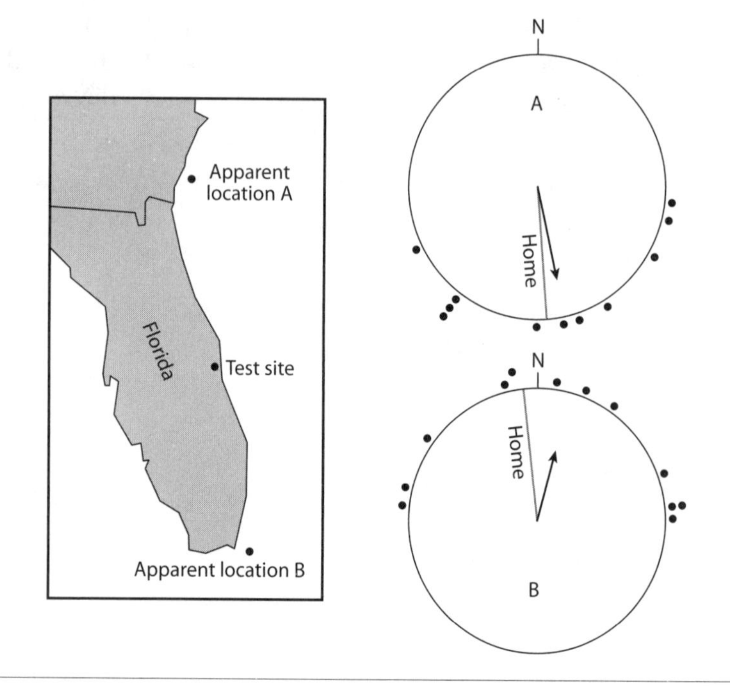

Virtual displacement of turtles. Older turtles captured on their feeding grounds were tested in a swimming tank, where they were presented with the magnetic field parameters characteristic of sites 210 miles north or south of their actual location. Those with an apparent northern displacement swam south; those in southerly magnetic conditions oriented north.

sity of 35,000 nT. Both groups responded appropriately for their longitude, with the ones apparently near Puerto Rico swimming NE toward the Sargasso Sea, whereas the turtles seeming to be off Cape Verde swam SW along the straight western flow of the gyre (groups H and F, respectively, in the gyre figure).

Although it's reasonable to conclude that the map sense can operate with nothing more than magnetic information, this does not mean that animals are ignoring other potentially useful sources of data. We have seen many instances in which local memory, inertial, and olfactory cues have been at work. But as a global-scale strategy that can provide (indirectly) both latitude and longitude, the magnetic GPS is far and away the most satisfactory explanation. Exactly how different species process the information—the relative roles of learning versus innate coordinates, or gradients versus simple intensity and inclination measurements, for instance—remains to be discovered.

■ A Human Map?

Soon after the original magnetic map hypothesis was formulated, British biologist Robin Baker published a startling paper documenting apparent human homing. He reported that blindfolded students taken by bus along a somewhat indirect route to a site 12 miles away could point back toward home; in other tests they could name the return direction. Moreover, if students wore magnets on their foreheads, their ability to infer their displacement was much reduced. Apparently humans must either have a map sense or excellent inertial navigation, and this ability must depend on a magnetic compass, a magnetic map, or both. The opportunities for follow-up work on this breakthrough were enormous.

As we looked at pictures of the test, we noticed that the students were only wearing sleep blindfolds. We consulted a local celebrity, James "The Amazing" Randi (known at the time for debunking a variety of paranormal performances). Randi had once successfully navigated an obstacle-strewn parking lot while driving blindfolded, so he knew something about defeating sensory-deprivation devices. While he assured us that nothing less than pizza dough covered by aluminum foil would provide absolute certainty, he helped design a loose-fitting, two-layer blindfold that seemed to cut out all light. We noticed too that the enormous windows of Baker's bus were uncovered, allowing sunlight to pour in and perhaps act as a beacon—or, given that the tests were run in often-overcast England, perhaps not. We decided to cover our bus windows with foil. Though Baker described his route as "circu-

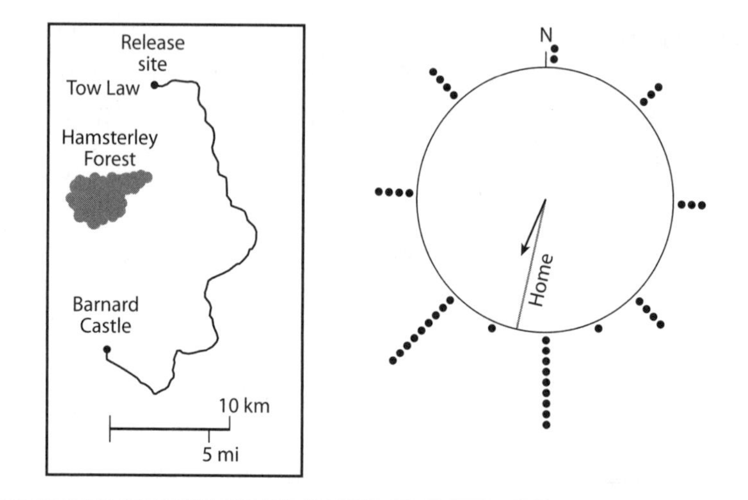

Original human-orientation test. Blindfolded students were taken by bus to a site roughly 12 miles north and asked to name the homeward direction. More than half of the students were accurate within 35°. Individuals wearing magnets were less precise.

itous," it seemed too linear to rule out inertial navigation; we decided on a less direct trip. Finally, students in his test knew if they were wearing magnets, and probably understood the expectation that this might interfere with their orientation. We decided that our controls would wear brass weights, avoiding any sort of placebo effect. (A few students nevertheless worked this out by placing their foreheads against metal trim in the bus, noting whether the object stuck or not.)

Our initial tests showed no evidence of human orientation. We then tried again, this time without magnets, using 13 different sites averaging 12 miles away. There was no homeward orientation at any of them.

Results of subsequent studies in the United Kingdom and Australia have been similarly negative. In common with our fellow primates, there is no good reason to believe that we have either a

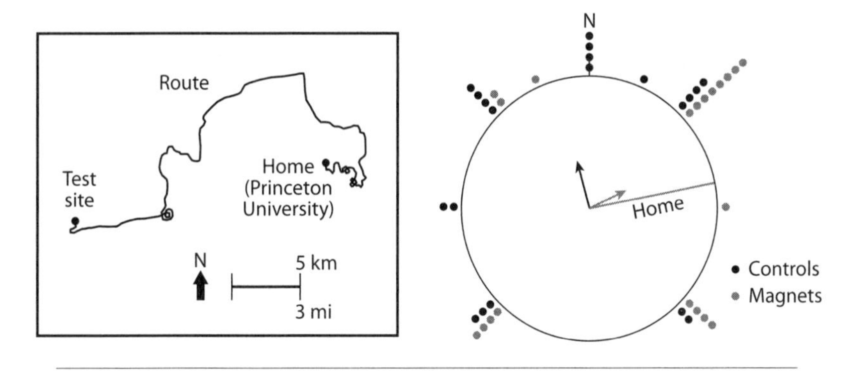

Princeton experiment. Students were taken by bus to a location about 11 miles west and asked to indicate the homeward direction.

map sense or a magnetic compass. We are well advised to rely on technology for these tasks.

The mental clocks and calendars of migratory species combine with their maps and compasses to give them an amazing ability to move about the planet. Despite the huge costs and risks of long distance travel they are able to exploit annual changes in weather and food. But the intervention of humans threatens to turn these adaptive redeployments into journeys toward oblivion. Is natural selection up to this challenge? Is there anything we can do to mitigate the likely damage? We will look at the future of migration in the next chapter.

Chapter 8

Migration and the Future: Conservation and Extinction

Why do animals migrate? For the southern right whales we studied as graduate students, the answer is clear; during the winter the Antarctic ice sheet expands, covering the summer feeding grounds. Though the rich concentration of krill is still there, air-breathing mammals can no longer safely feed. And because whales are warm blooded, lingering in frigid water waiting for the return of spring is metabolically expensive. Instead, the right whales make annual pilgrimages to traditional coastal sites in the temperate zone 2000 miles to the north.

For the group of about 60 whales we worked with, the winter refuge is Golfo San José on the Patagonian coast of Argentina. There they give birth and mate, relatively safe in protected bays from families of killer whales on the hunt. Like all migrants, right whales are sensitive to habitat and climate changes, as well as to human activity. These gentle creatures have been hunted nearly to extinction, and the remainder have been exiled from bays with shipping or industrial development. Rising ocean temperatures are pushing surviving populations of whales, as well as other sea creatures, to alter the timing of their age-old migrations. Given that 12% of bird species (including 45% of seabirds), most large whales, and all sea turtles are endangered, we must wonder how well migrants deal with these challenges, and what their long-term prospects look like.

■ Evolution of Migration

Migration evolved because its benefits outweigh the associated costs. The logic of natural selection dictates that a bird that breeds in the Arctic and overwinters in the tropics must be producing more surviving offspring on average, despite the costs in time and energy of flying hundreds or thousands of miles north to breed. Otherwise such a expensive system would not persist. In fact long-distance migrants have been doing especially well in the recent past; although they produce smaller clutch sizes and fewer broods, they have on average the same number of successful young as residents and short-distance migrants. Predation and starvation of fledglings in the tropics and temperate zone must be very high compared to the situation farther north, where the ephemeral bursts of spring growth make food briefly plentiful, while the cruel winters severely cull or eliminate predator populations. And not having to work as hard at parenting, adults live longer.

Though habitat and climate change are the focus of conservationist concern for migrants, these are precisely the factors that appear to have selected for migration in the first place. Change, of course, is inevitable; to accommodate it an animal can alter its mix of genes or its location (or, most often, both). The degree of change animals have faced through evolutionary history (600 million years [MY] for vertebrates, 200 MY for birds) is enormous. Average global temperatures have ranged from below freezing to above 100°F—and these are just averages. During ice ages the poles were even colder, while the tropics experienced still hotter weather during some of the interglacial interludes. During periods when all the polar ice has melted ocean levels have been 250 feet higher than they are at present; in times of global freezing the sea level has been almost 450 feet lower (though this has not happened for 700 MY). Vegetation has tracked this change in climate as the boundaries between tropical, temperate, and boreal forests moved thousands of miles in latitude. Animals had to relocate or go extinct. One obvious solution was to migrate, which allows animals to take advantage of the annual global differences in weather.

But it's not quite that simple. The usual picture of an all-or-nothing, species-wide choice of strategy is misleading; most species of birds, for instance, have a continuum of options. Some of the finches we see in the summer in Princeton have flown north

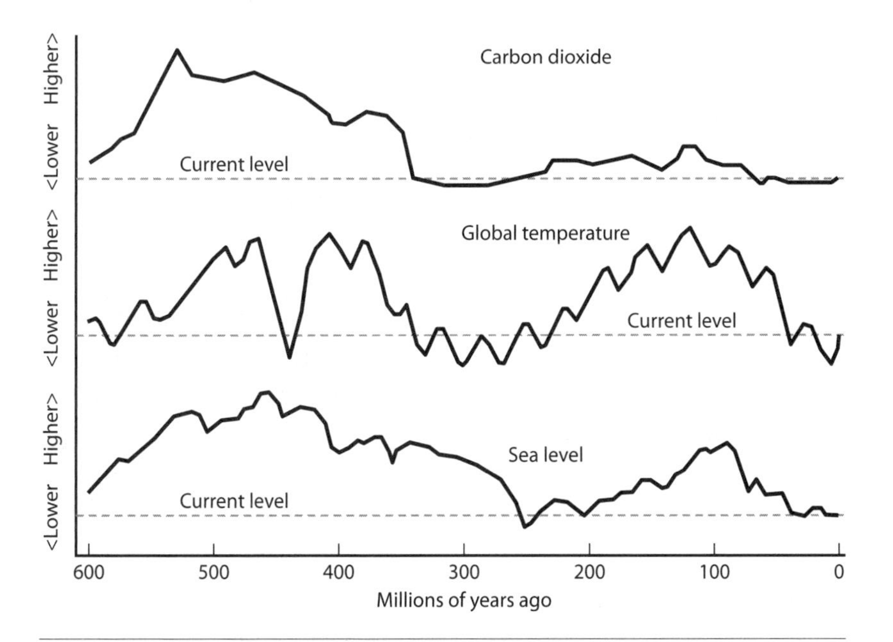

Climate variability in the past. Over the 600 million years that vertebrates have existed the earth has undergone major swings in climate and other characteristics important to migrants. Although in the simplest case carbon dioxide levels should control temperature, and temperature should control sea level (through the accumulation or melting of ice), note that the detailed correlation is not particularly strong. By historical standards, the planet has relatively little carbon dioxide in the atmosphere at the moment, is much cooler than average, and is experiencing a period of relatively low sea level.

from winter quarters on the Gulf of Mexico, but others have spent the cold months locally; some (not many, but their numbers are increasing) vacationed in between. In fact, the majority of bird species (55–60%) have a mix of individuals that have adopted either the resident strategy or the migratory alternative. Even among the permanent residents, though, migratory restlessness is evident in the spring and fall as some inner programming prompts the animals to prepare for journeys they will never undertake.

Birds within a given species typically differ in many relevant traits. There may be variations in the timing of spring and autumn migration (or the corresponding restlessness). The default direction for a young bird's first fall journey may vary with the population. There may be differences in the preferred distance (or amount of time) to fly, the ideal stopping latitude both for spring and fall, the preferred speed, and the number or duration, or both, of stopovers. The weighting of any cues that identify a suitable place to end the journey or pause along the way also may vary between populations and individuals. Each of these parameters is to a large extent genetic. But equally innate is a specific degree of *phenotypic plasticity*, a capacity for day-to-day change that confers an ability to alter behavior in response to current or past contingencies such as a bout of unusually warm weather or unfavorable winds.

Thus two sparrows under the same conditions may have different departure dates. A cross between them will produce offspring with an intermediate date, showing that this departure time is genetically encoded. And yet both will respond to a week of unusually warm weather in the late winter by starting their northward migration a day or two earlier, employing their neural weather algorithm to produce an adaptive one-time phenotypic adjustment.

Phenotypic fine tuning is generally stronger in resident populations, even compared to migrants of the same species. It is typically greater in short-distance migrants (seasonal journeys less than 200 miles) than in long-range species. But it can differ between groups; great tits in England adjust their egg-laying dates over a relatively wide range to match spring temperatures, while those in the Netherlands seem on average to ignore the weather. Even in the Dutch population, however, there is considerable genetic variation for the degree of plasticity, variation upon which selection can operate should a systematic advantage of one strategy over the other lead to greater relative reproductive fitness.

For evolution to occur though natural selection, migratory parameters must vary among individuals. Variation has to be heritable, and different variants need to have differential reproductive success. The variation in these traits is clear in migrating birds, but the *degree* of variation differs dramatically between species. A narrow range of genetic alternatives typically reflects a specialist species that has perfected an optimal strategy from which it barely strays. Low variation also is characteristic of small populations or species that have been through a "bottleneck" event in which population size (and the corresponding genetic variability) was severely pruned. The North American whooping crane population, for instance, numbered only 21 individuals in 1941, but has recovered to about 400 wild birds after intense conservation efforts. On average, we would expect species with low variability or low phenotypic plasticity to do less well in the face of environmental change.

Selection affects the spread and centering of this variability in two major ways. Most commonly, normalizing selection trims the extremes of the distribution. Cliff swallows are a good example of normalizing selection. In Nebraska these birds arrive over a two-week period in the spring; the date of their first arrival has drifted earlier by about three days per decade in recent years. A severe cold snap in 1996, however, killed off the early arrivers. Because this trait is genetic, no birds arrived early in the subsequent year. Selection occurs at the other end of the flight-time distribution as well: in good years late-arriving swallows find the best nesting sites and most-fit mates already taken, and thus experience severely re-

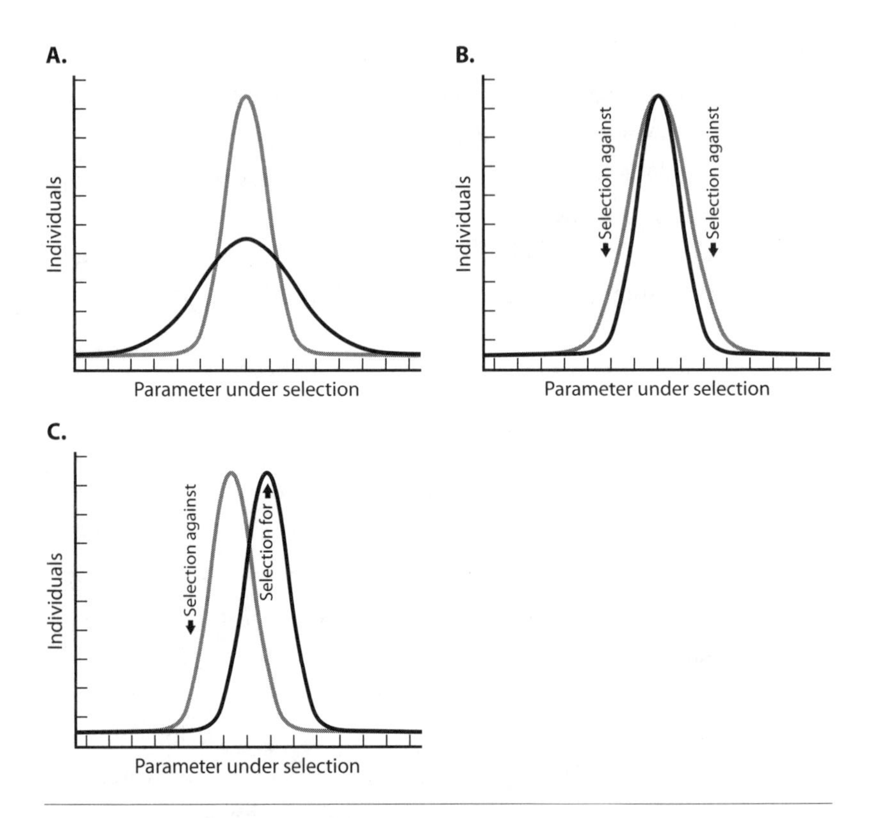

Evolution of traits. (A) Species differ with regard to the amount of variation in the genome for parameters such as starting date, speed of travel, vector of first fall migration, and so on. Specialists and species with a small population size tend to have less variation (the narrow curve). (B) Most selection is normalizing, cropping off the ends of the distribution, thus narrowing the curve. (C) As conditions change, selection generally becomes directional, operating against one extreme or in favor of the other end, or both, shifting the distribution.

duced reproductive success. Bad weather typically removes less symmetrical individuals as well (which are presumably less fit), as well as unusually large and small members of the population. But if selection systematically favors one extreme or punishes the other (or, quite often, both), the mean value of the parameter will move in a consistent direction.

Because the start of the growing season above 45°N latitude has advanced 12–19 days in the last 50 years, both phenotypic compensation and selection for earlier or faster migration seem likely to be taking place. And in fact a variety of species have already accommodated the warming trend in higher northern latitudes. In the Finger Lakes region of New York State, for example, 26 of the 34 species of short-distance migrants breeding in the area arrived significantly earlier in the last half of the 20th century compared with the first half—all by at least a week, but in several cases by a month or more. The pattern is the same in Europe; reed warblers, for instance, have advanced their egg laying by about 20 days over the last 40 years. The capacity for change is even greater than this. Breeding experiments that push selection pressure artificially high can move up the departure date by a week or delay it by a fortnight in just two years without any need for phenotypic plasticity to accelerate the shift. Indeed, both breeding results and analyses of related populations with different migratory patterns show that all components of the navigational repertoire seem quick to respond to selection.

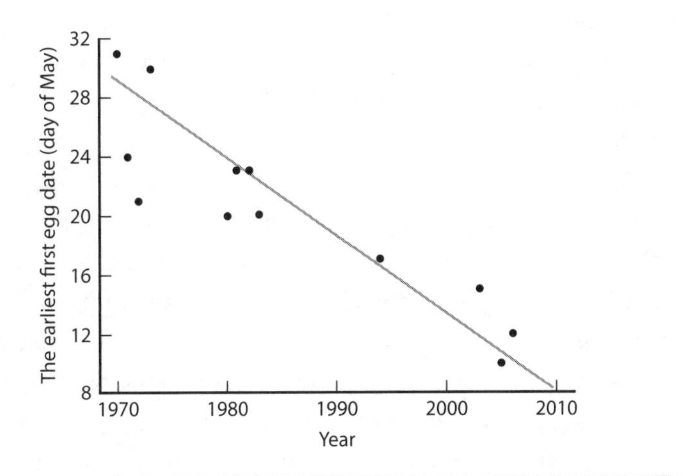

Laying dates for reed warblers. Reed warblers begin nesting almost three weeks earlier now than they did 40 years ago.

When three or more genes are at work in controlling the expression of a trait (which is the usual situation for any but the simplest factors), much of the genetic variation is normally hidden by the additive way in which the genes combine with one another. If, for instance, you were to flip a coin six times, you would get all heads only once in 64 tries; similarly, the most extreme combinations of genes in offspring are rarely seen. It is this reservoir of covert variability that allows selection to work so quickly.

The house finch population in North America is a dramatic example of hidden migratory potential. The natural range of this species until the 20th century was the American Southwest and western Mexico, where only 2–3% are migratory; the rest were year-round residents. In the 1940s California house finches were widely marketed in New York City and on Long Island as cage birds (so-called "Hollywood" finches). Apparently this practice ended with the threat of prosecution, though accounts differ as to which law was invoked. Most of the caged birds were released into the wild and can now be found throughout the United States, where they have largely displaced the native purple finch. The interesting thing is that in the harsh winters of the eastern and midwestern United States, 40–80% of the finches (depending on the exact location) are now migratory. Selection strongly favored the rare combination of genes that impelled the birds to fly south for the winter, and the species responded quickly.

Although directional selection is the most common agent of evolution, another powerful mechanism is a combination of chance and inbreeding. Suppose some outlier in the distribution—individuals with a highly unusual innate set of initial migration bearings or flight distances—were to stumble upon a favorable breeding habitat. They would enjoy enhanced reproductive success, and moreover mainly breed with other animals sharing the same unusual genetic proclivities. An entirely new migratory population could then evolve though what is commonly called a *founder effect*.

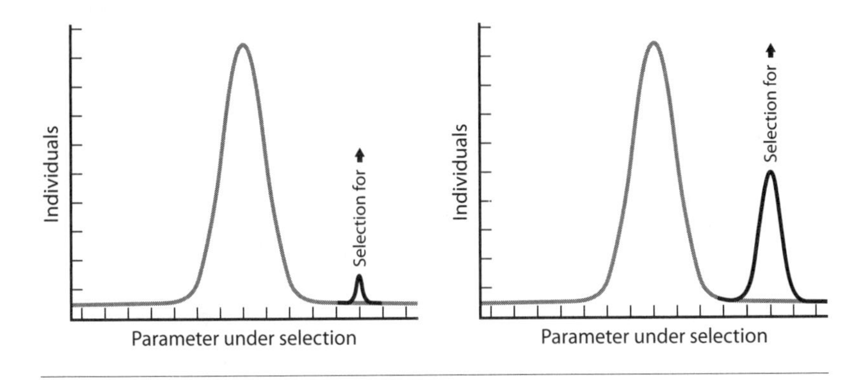

Evolution of a new population. Selection need not operate gradually. If a few individuals in one tail of the parameter distribution discover a new, highly favorable habitat, and if as a consequence they are reproductively isolated from the rest of the species, they may inbreed and create a genetically distinct population.

In fact this pattern of navigational innovation, essential for the spread of migratory species, has been observed or inferred numerous times. One particularly clear case, investigated in remarkable detail by Peter Berthold and his colleagues at the Max Planck Institute for Ornithology, involves the blackcap, a species of warbler. Blackcaps breed across northern Europe and typically winter in Spain. Each population has an average innate default departure vector appropriate to its summer location, and an equally innate sense of how far to travel. Within the population different individuals vary to some degree in their distance and direction proclivities. At least for flight duration the genetic basis of the variation is now understood; it depends on the number of two-base repeats in a gene unmemorably called *ADCYAPI*, which affects circadian rhythms and energy use.

Blackcaps hardly ever overwintered in the United Kingdom prior to 1950, even though Great Britain is on the route for the Norwegian population. But beginning about half a century ago, a group of warblers started spending the cold months in England and Wales. Cage-reared birds from Wales tell us that the innate first-fall vector is aimed west rather than SW (the typical direction for most other populations of these warblers). This suggested that the spring breeding area must be in Germany or Austria. Genetic analyses show that, in fact, nearly 10% of the blackcaps in these regions are now traveling to the United Kingdom for the winter. Crossing the two populations produces birds that prefer to fly WSW (into oblivion in the North Atlantic).

The likely scenario is that a few warblers from the western extreme of the vector distribution and the short end of the range of flight duration variation found the United Kingdom, did well, and returned to breed. They maintain the requisite genetic isolation by

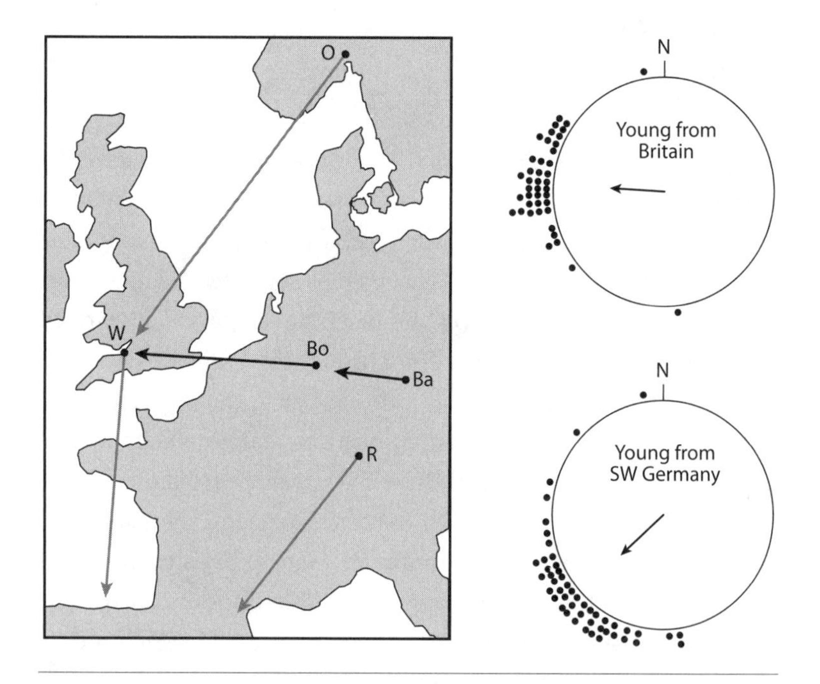

Rapid evolution of a new migration strategy in warblers. Blackcaps generally migrate SW in the fall and typically winter in Spain. One population (O) migrates through the United Kingdom en route. Since 1950 a population breeding in Germany has begun migrating west and wintering in Wales and southern England. The vector preference is strictly genetic: the bearings shown here represent hand-reared birds in their first autumn.

virtue of arriving back at the breeding ground early (a consequence of their much reduced route), completing their courtship and pairing before the Spanish contingents return.

■ Climate and Habitat Change

The threat of climate change and the ensuing change in global habitats have been concerns for decades. Discussion of their catastrophic potential reached a peak in the early 1970s when the National Academy of Sciences reported a "widely held consensus" that a major change was inevitable, and might already have begun. The overhyped worry then, however, was a repeat of the ice age of 20,000 years ago, which buried New York City under a thousand feet of glacial mass. This experience suggests that current furor over global warming merits some degree of initial skepticism. The situation is not helped by the political overtones and intellectual intolerance the debate has taken on, nor by the relatively uninformed use of often questionable or irrelevant data. For instance, hardly anything in the well-meaning film starring Al Gore, *An Inconvenient Truth* (2006), is wholly true.

Fortunately, there are two clear sources of data unaffected by political predispositions. The first we've already encountered: birds are migrating and nesting sooner, responding to the earlier growing season. Obviously these animals believe the planet is warming, and are betting their lives on this conclusion. The second is the ocean. Sea level changes in consequence of two main factors. First, the volume of water increases with temperature as a simple result of thermal expansion; second, it rises when terrestrial ice melts and the water makes its way to the ocean. Both phenomena are obvious consequences of global warming, though expansion is the dominant process at the moment. The oceans are, in essence, a huge, volume-based thermometer, and sea level has been increasing steadily. Whether this is mainly the result of human activity

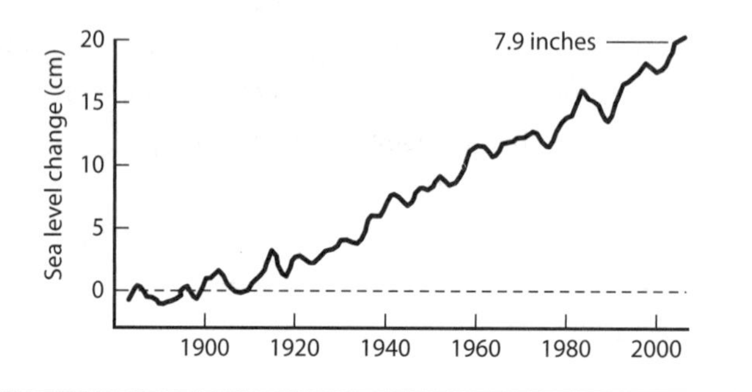

Recent trends in sea level. Sea level has risen about eight inches in the last century, primarily as a result of the thermal expansion of increasingly warm ocean water.

raises another set of loaded questions; either way it's potentially a huge challenge for the earth's flora and fauna.

What exactly can long-distance migrants expect from this change in average global temperature? The earth's great biomes are on the move. The tundra, which as we have seen supports large populations of migrants with its boom-and-bust cycles of growth, remains characteristically treeless because its spongy surface retains water that cannot drain away through the underlying permafrost. As temperatures rise the permafrost recedes, opening the moss-dominated tundra to the incursion of forest. Assuming no abatement of warming, even conservative estimates put the treeline boundary between boreal forest and tundra up to 250 miles farther north within the next two centuries, with as much as 40% of the tundra changing into evergreen-dominated taiga in the process. Much of the current taiga, in turn, will be encroached upon by temperate forest flora, which will eventually be under pressure at its southern boundary from tropical species.

Because it is the tundra that supports so many of the long-distance migrants, and particularly birds, it follows that although the growing season will lengthen, the habitat area will shrink enormously. Unless the extended summer allows for a round of second nesting, it's hard to see how the earlier spring will benefit the birds.

In any event, most long-distance migrants seem to be advancing the start of their spring migration departures only minimally. The tropics, where they spend the winter, provide few cues to signal the onset of spring thousands of miles to the north. Long-distance migrants must depend on their innate programming to tell them when to leave, and thus miss out on the sort of genetic variability and phenotypic response that power the adaptability we see in short-range migrants. On the other hand, records show that long-distance migrants are traveling faster on their way to high latitudes, probably because they do not need to wait for their climate-dependent food supply en route to catch up. For shorebirds there is another potential difficulty: the rising sea level will probably consume wetlands faster than succession can create new ones suitable for foraging and breeding. Swamp and marshland take a long time to develop into the richly productive habitats the birds depend on.

For residents and short-distance migrants the picture is slightly different. For one thing, in the Northern Hemisphere the boundary between residents and migrants has been moving north, having shifted more than 25 miles poleward in the past 30 years. Correspondingly, short-distance migrants like blackcaps are tending to truncate their southward journeys in the fall, overwintering farther north as well. And the proportion of birds rearing a second or even a third brood in the spring and summer has increased sharply. Here in Princeton a second nesting by house wrens was unusual 40 years ago; today even a third round of reproduction is commonplace. In the United Kingdom some overachieving sparrows are now rearing four broods each year. The greater phenotypic plasticity and range of hidden genetic variation in residents and short-distance migrants enables them to accommodate the change, and even profit by it. It remains to be seen whether this innate adaptability is sufficient to keep pace with the relatively rapid rate of climate change currently under way.

But while many species have the inherent capacity to accommodate climate change, the more serious threat for most is habitat destruction. More than 40% of endangered birds are in that category because of habitat loss somewhere in their range. For birds adapted to reproduce in grasslands, the worldwide loss of 25% of this habitat to crops is a serious blow. Birds overwintering in the tropics face a 5% loss of rain forest per decade. Depending on your preferred definition, forests worldwide cover only 50–65% of their former expanse, removing not only breeding habitat essential for some residents and migrants, but places for long-distance migrants to stop and feed on the way to their more distant targets. Even where forests have declined less they are increasingly fragmented, a process that creates a distinctly different forest-edge habitat at the expense of a mid- and deep-forest ecology, areas that support distinctly different species. Predators and nest parasites are especially common on forest edges, and human development is parceling what were vast sweeps of deep secure forest environment into small patches, surrounded by perilous edge.

By far the biggest victims of habitat destruction are the shorebirds: the rich wetlands environment that is life to them and so many other creatures is valuable coastal real estate to humans. The "reclamation" of marshes and mangroves, the damming of streams, and pollution of bays and rivers by cities has reduced shorebird habitat drastically.

While we have focused on avian migrants, the situation is equally challenging for other long-distance travelers. Monarch butterflies are steadily losing their relatively small wintering habitat in the mountains west of Mexico City to illegal logging. Sea turtles are losing their nesting habitats on beaches to development, and risk being caught by long-line fishing or trawling while at sea. Salmon are losing their eggs and fry to sediment-heavy runoff from forest logging while adults are exiled from their native streams by dams and introduced sport fish. Wild salmon are overfished, and are increasingly diseased from parasites picked up from fish farms they pass on their way to the sea. At home or on the move, the combination of climate change and habitat loss is threatening migrants to an unprecedented degree. In the disturbing words of conservationist David Wilcove in his prescient book *No Way Home* (2008), a migrating animal "travels without any knowledge of what may have happened to its breeding grounds, its wintering grounds, or any of the places in between since the last time it made the journey. . . . Migration is an act of faith after all, a hardwired belief that there is somewhere to go to and a way to get back."

Against this set of challenges we are fortunate to have an increasing understanding of the ways animals navigate. Whooping cranes are recovering in large part because we now know about their strip-map approach to learning and remembering their migratory route. As a result young whoopers can be reared in an incubator anywhere on earth, imprinted on an ultralight aircraft, and then led along an arbitrary route complete with stopovers to a protected wintering ground. They, in their turn, will lead the next generation through the same journey. Similarly, the "devil bird" cahow is being rescued from the brink of extinction because conservationists on Bermuda understand the phenomenon of site imprinting. Cahow chicks, abandoned by their parents, emerge from their burrows and memorize the magnetic parameters of their nest site on the night they fledge. Thus they can be moved to a safer nesting site any time up to the day before they take wing, and will return to that same spot when they are ready to breed five years later.

But things are not always this straightforward. For instance, conservationists would like to take advantage of magnetic nest-site imprinting in sea turtles by moving the eggs to new beaches before hatching, thus reestablishing extinct populations. But this needs to be done quite late in the developmental period: sex in turtles depends on incubation temperature, and is calibrated by stabilizing selection to the conditions of the natal beach to yield a 50:50 mix. Thus when biologists attempted to reintroduce green sea turtles to Bermuda from the Caribbean, the lower sand temperatures on the Isles of Devils resulted in thousands of all-male hatchlings—turtles that we observe returning each year in a futile search for mates.

In addition to the power our understanding of strip maps and true maps confers, the beacon-based system of salmon allows conservationists to reintroduce these migrants to streams they have "forgotten," or to establish new populations in promising, unpolluted locales. The process begins with imprinting fry raised in hatcheries with an artificial odor on the first day of the smolt stage, then releasing them into the brackish waters near the mouth of the river system in question to memorize the map coordinates. The target stream then needs to be baited with the artificial odor the appropriate number of years later to guide returning adults. Afterward the artificial odor is no longer needed; the natural aromas of the rivulet will be memorized by the new generation of smolts six months or more later before they head downstream.

Increasingly researchers are making use of technological solutions originally developed to solve the problems posed by our own innate navigational shortcomings—the problems that doomed the *San Antonio*'s crew. Miniaturized GPS trackers, for instance, in addition to getting clueless motorists to their destinations, can reconstruct the journeys of many species of migratory animals, revealing how their onboard compass and map senses are guiding their travels as well as their choice of stopovers and termination point. This information provides insights into migratory pathways and the cues animals use, both in terms of their sensory equipment and neural programming.

Often these results tell conservationists what *won't* work. Recall that most migrating birds (waterfowl excepted) set off that first autumn along an innately specified vector. Only genetic variation and selection can do much to alter their instinctive headlong rush into the unknown. Given that we cannot put every endangered migrant on emergency conservational life support, some sort of triage is inevitable. For many species, variation and selection provide the best chance these victims of human population growth have for long-term survival.

A wider understanding and appreciation among the peoples of the several independent nations crossed by a migrant's journey will be essential in preserving the habitats these animals need, habitats that may only be used a few weeks each year. Expecting people to stop using the planet and its resources—to forego growing needed crops, for instance, to make life easier for transient grassland species—is useless. Preventing short-term climate change is an equally fantastic dream. But seeking international cooperation in highly targeted and well-financed ways to accommodate migrating species is in many cases a realistic goal, while we work out the ways to shape or at least ameliorate our planet's destiny.

We understand enough—or perhaps just know enough to realize how to fill in the gaps in our understanding—to help at least some of these species survive. What's mainly lacking is the irrational, optimistic zeal necessary to make conservation work—irrational in the sense that we must assume with no very good reason that we can decode mysteries and use those discoveries to solve real-world problems. We need to tell the fascinating story of navigation and migration to those in and along the travel corridors, communicating the awe and wonder that fuels the research of most serious biologists.

This should not be impossible. Whether it's their ability to judge time and distance, use vectors and beacons, create cognitive maps, take compass bearings from cues indecipherable to us, or draw on an inborn map sense to position themselves on the planet, there are enough intriguing mysteries in animal navigation to engage the imagination and creative energies of new legions of conservationists around the globe.

Bibliography

Chapter 1

Adler, J. (1976). The sensing of chemicals by bacteria. *Scientific American* 234 (4), 40–47.

Blakemore, R. P. (1975). Magnetotactic bacteria. *Science* 190, 377–379.

Bochdansky, A. B., and Bollens, S. M. (2004). Relevant scales in zooplankton ecology: Distribution, feeding, and reproduction of the copepod *Acartia hudsonica* in response to thin layers of the diatom *Skeletonema costatum*. *Limnology and Oceanography* 49, 625–636.

Farkas, S. R., and Shorey, H. H. (1972). Chemical trail following by flying insects. *Science* 178, 67–68.

Fraenkel, G. S., and Gunn, D. L. (1940). *The Orientation of Animals: Kineses, Taxes, and Compass Reactions* (Oxford, United Kingdom: Clarendon).

Frederiksen, R., Wcislo, W. T., and Warrant, E. J. (2008). Visual reliability and information rate in the retina of a nocturnal bee. *Current Biology* 18, 349–353.

Gould, J. L. (1982). *Ethology: The Mechanisms and Evolution of Behavior* (New York, NY: W. W. Norton).

Gould, J. L., and Gould, C. G. (1995). *The Honey Bee*, 2nd ed. (New York, NY: W. H. Freeman).

Haney, J. F. (1988). Diel patterns of zooplankton behavior. *Bulletin of Marine Sciences* 43, 583–603.

Kennedy, J. S., Ludlow, A. R., and Sanders, C. J. (1980). Guidance system used in moth sex attraction. *Nature* 288, 475–477.

Kennedy, J. S., and Marsh, D. (1974). Pheromone-related anemotaxis in flying moths. *Science* 184, 999–1001.

Lohmann, K. J., Cain, S. D., Dodge, S. A., and Lohmann, C. M. F. (2001). Regional magnetic fields as navigational markers for sea turtles. *Science* 294, 364–366.

Lohmann, K. J., and Lohmann, C. M. F. (1996). Orientation and open-sea navigation in sea turtles. *Journal of Experimental Biology* 199, 73–81.

Maloney, E. S. (1978). *Dutton's Navigation and Piloting* (Annapolis, MD: Naval Institute Press).

Marx, R. F. (1987). *Shipwrecks of the Western Hemisphere, 1492–1825* (New York, NY: Dover).

May, W. E., and Holder, L. (1973). *A History of Marine Navigation* (Henley-on-Thames, United Kingdom: G. T. Foulis).

Williams, J. E. D. (1992). *From Sails to Satellites: The Origin and Development of Navigational Science* (Oxford, United Kingdom: Oxford University Press).

Chapter 2

Frederiksen, R., Wcislo, W. T., and Warrant, E. J. (2008). Visual reliability and information rate in the retina of a nocturnal bee. *Current Biology* 18, 349–353.

Gould, J. L. (1988). Resolution of pattern learning in honey bees. *Journal of Insect Behaviour* 1, 225–233.

Gould, J. L., and Gould, C. G. (1995). *The Honey Bee*, 2nd ed. (New York, NY: W. H. Freeman).

Gwinner, E., and Wiltschko, W. (1978). Endogenously controlled changes in migratory direction of the garden warbler. *Journal of Comparative Physiology* 125, 267–273.

Lincoln, F. C., Peterson, S. R., and Zimmerman, J. L. (1998). *Migration of Birds* (Washington, DC: US Department of the Interior, US Fish and Wildlife Service, Circular 16).

Maloney, E. S. (1978). *Dutton's Navigation and Piloting* (Annapolis, MD: Naval Institute Press).

May, W. E., and Holder, L. (1973). *A History of Marine Navigation* (Henley-on-Thames, United Kingdom: G. T. Foulis).

Riley, J. R., Reynolds, D. R., Smith, A. D., et al. (1999). Compensation for wind drift by bumble bees. *Nature* 400, 126.

Williams, J. E. D. (1992). *From Sails to Satellites: The Origin and Development of Navigational Science* (Oxford, United Kingdom: Oxford University Press).

Chapter 3

Arenz, A., Silver, R. A., Schaefer, A. T., and Margrie, T. W. (2008). The contribution of single synapses to sensory representation in vivo. *Science* 321, 977–980.

Aschoff, J. (1965). Circadian rhythms in man. *Science* 148, 1427–1432.

Bentley, G. E. (2008). Biological timing. *Current Biology* 18, R736–R738.

Bolsvert, M. J., and Sherry, D. F. (2006). Interval timing by an invertebrate, the bumble bee. *Current Biology* 16, 1636–1640.

Collett, M., Collett, T., and Srinivasan, M. V. (2006). Insect navigation: Measuring travel distance across ground and through air. *Current Biology* 20, R887–R890.

Collett, T. S. (1993). Route following and the retrieval of memories in insects. *Comparative Biochemistry and Physiolology A: Comparative Physiology* 104, 709–716.

Cordes, S., and Gallistel, C. R. (2008). Intact interval timing in circadian CLOCK mutants. *Brain Research* 1227, 120–127.

Dacke, M., and Srinivasan, M. V. (2008). Two odometers in honey bees? *Journal of Experimental Biology* 211, 3281–3286.

Durgin, F. H., Akagi, M., and Gallistel, C. R. (2008). The precision of locomotor odometry in humans. *Experimental Brain Research* 193, 429–436.

Esch, H., and Burns, J. (1996). Distance estimation by foraging honeybees. *Journal of Experimental Biology* 199, 155–162.

Floreano, D., and Zufferey, J.-C. (2010). Insect vision: A few tricks to regulate flight altitude. *Current Biology* 20, R847–R849.

Frederiksen, R., Wcislo, W. T., and Warrant, E. J. (2008). Visual reliability and information rate in the retina of a nocturnal bee. *Current Biology* 18, 349–353.

Gallistel, C. R., and Gibbon, J. (2000). Time, rate, and conditioning. *Psychological Review* 107, 289–344.

Gallistel, C. R., King, A., and McDonald, R. (2004). Sources of variability and systematic error in mouse timing behavior. *Journal of Experimental Psychology: Animal Behavior Processes* 30, 3–16.

Gould, J. L. (1980). Sun compensation by bees. *Science* 207, 545–547 (1980).

Gould, J. L. (1988). Resolution of pattern learning in honey bees. *Journal of Insect Behaviour* 1, 225–233.

Gould, J. L., and Gould, C. G. (1995). *The Honey Bee*, 2nd ed. (New York, NY: W. H. Freeman).

Gross, H. J., Pahl, M., Si, A., et al. (2009). Number-based visual generalization in the honey bee. *PLoS ONE* 4 (e4263), 1–9.

Gwinner, E., and Wiltschko, W. (1978). Endogenously controlled changes in migratory direction of the garden warbler. *Journal of Comparative Physiology* 125, 267–273.

Hazlerigg, D., and Loudon, A. (2008). New insights into ancient seasonal timers. *Current Biology* 18, R795–R804.

Kavanau, J. L. (1969). Behaviour of captive white-footed mice. In E. R. Willems and H. L. Raush (Eds.), *Naturalistic Viewpoints in Psychology* (New York, NY: Holt, Rinehart, Winston), pp. 221–270.

Layne, J. E., Barnes, W. J. P., and Duncan, L. M. J. (2003). Mechanisms of homing in the fiddler crab *Uca rapax*. *Journal of Experimental Biology* 206, 4425–4442.

Lent, D. D., Graham, P., and Collett, T. S. (2009). A motor component to the memories of habitual foraging routes in wood ants? *Current Biology* 19, 115–121.

Lindauer, M. (1977). Recent advances in the learning and orientation of honey bees. In *Proceedings of the Fifteenth International Congress of Entomology* (College Park, MD: Entomological Society of America), pp. 450–460.

Maloney, E. S. (1978). *Dutton's Navigation and Piloting* (Annapolis, MD: Naval Institute Press), p. 14.

May, W. E., and Holder, L. (1973). *A History of Marine Navigation* (Henley-on-Thames: G. T. Foulis).

Meder, E. (1958). Über die Einberechnung der Sonnenwanderung bei der Orientierung der Honigbiene. *Zeitschrift für Vergleichende Physiologie* 40, 610–641.

Merlin, C., Gegear, R. J., and Reppert, S. M. (2009). Antennal circadian clocks coordinate sun compass orientation in migratory monarch butterflies. *Science* 325, 1700–1704.

Mittelstaedt, M., and Mittelstaedt, H. (2001). Idiothetic navigation in humans: Estimation of path lengths. *Experimental Brain Research* 139, 318–332.

Moore, D., and Rankin, M. A. (1983). Diurnal changes in the accuracy of the honey bee foraging rhythm. *Biological Bulletin* 164, 471–482.

Pengelley, E. T. (Ed.). (1974). *Circannual Clocks: Annual Biological Rhythms* (New York, NY: Academic Press).

Renner, M. (1957). Neue Versuche über den Zeitsinn der Honigbienen. *Zeitschrift für Vergleichende Physiologie* 40, 85–118.

Riley, J. R., Reynolds, D. R., Smith, A. D., et al. (1999). Compensation for wind drift by bumble bees. *Nature* 400, 126.

Shafer, M., and Wehner, R. (1993). Loading does not affect measurement of walking distance in desert ants, *Cataglyphis fortis*. *Verhandlungen der Deutschen Zoologen Gesellschaft* 86, 270.

Shaw, A. D., Lee, S., and Dickinson, M. H. (2010). Visual control of altitude in flying *Drosophila*. *Current Biology* 20, 1550–1556.

Shine, H. (1996). Optokinetic speed control and estimation of travel distance in walking honeybees. *Journal of Comparative Physiology A: Neuroethology, Sensory, Neural, and Behavioral Physiology* 179, 587–592.

Srinivasan, M. V., Zhang, S. W., and Bidwell, N. J. (1997). Visually mediated odometry in honey bees. *Journal of Experimental Biology* 200, 2513–2522.

von Frisch, K. (1967). *The Dance Language and Orientation of Bees* (Cambridge, MA: Harvard University Press).

Walls, M. L., and Layne, J. E. (2009). Direct evidence for distance measurement via flexible stride integration in the fiddler crab. *Current Biology* 19, 25–29.

Waterman, T. H. (1989). *Animal Navigation* (New York, NY: W. H. Freeman), p. 201.

Williams, J. E. D. (1992). *From Sails to Satellites: The Origin and Development of Navigational Science* (Oxford, United Kingdom: Oxford University Press).

Wittlinger, M., Wehner, R., and Wolf, H. (2006). The ant odometer: Stepping on stilts and stumps. *Science* 312, 1965–1967.

Chapter 4

Blakemore, R. P. (1975). Magnetotactic bacteria. *Science* 190, 377–379.

Brines, M. L., and Gould, J. L. (1979). Bees have rules. *Science* 206, 571–573.

Brines, M. L., and Gould, J. L. (1982). Skylight polarization patterns and animal orientation. *Journal of Experimental Biology* 96, 69–91.

Brower, L. P. (1996). Monarch butterfly orientation. *Journal of Experimental Biology* 199, 93–103.

Brun, R. (1914). *Die Raumorientierung der Ameisen* (Jena, Germany: Fischer).

Cardè, R. T. (2008). Insect migration: Do migrant moths know where they are heading? *Current Biology* 18, R472–R474.

Chapman, J. W., Reynolds, D. R., Mouritsen, H., et al. (2008). Wind direction and drift compensation optimize migratory pathways in a high-flying moth. *Current Biology* 18, 514–518.

Collett, T. S. (2008). Insect navigation: Visual panoramas and the sky compass. *Current Biology* 18, R1058–R1061.

Dacke, M., Byrne, M. J., Scholtz, C. H., and Warrant, E. J. (2004). Lunar orientation in a beetle. *Proceedings of the Royal Society of London B: Biological Sciences* 271, 361–365.

Dacke, M., Nilsson, D.-E., Scholtz, C. H., et al. (2003). Insect orientation to polarized moonlight. *Nature* 424, 33.

Dacke, M., Nilsson, D.-E., Warrant, E. J., et al. (1999). Built-in polarizers form part of a compass organ in spiders. *Nature* 401, 470–473.

Dalton, J. (1798). Extraordinary facts relating to the vision of colours: With observations. *Memoirs of the Literary and Philosophical Society of Manchester* 5, 28–45.

Diebel, C. E., Proksch, R., Green, C. R., et al. (2000). Magnetite defines a magnetoreceptor. *Nature* 406, 299–302.

Dyer, F. C. (1987). Memory and sun compensation by honey bees. *Journal of Comparative Physiology A: Neuroethology, Sensory, Neural, and Behavioral Physiology* 160, 621–633.

Dyer, F. C., and Dickinson, J. A. (1994). Development of sun compensation by honey bees. *Proceedings of the National Academy of Sciences* 91, 4471–4474.

Dyer, F. C., and Gould, J. L. (1981). Honey bee orientation: A backup system for cloudy days. *Science* 214, 1041–1042.

Dyer, F. C., and Gould, J. L. (1983). Honey bee navigation. *American Scientist* 71, 587–597.

Frederiksen, R., Wcislo, W. T., and Warrant, E. J. (2008). Visual reliability and information rate in the retina of a nocturnal bee. *Current Biology* 18, 349–353.

Gegear, R. J., Casselman, A., Waddell, S., and Reppert, S. M. (2008). Cryptochrome mediates light-dependent magnetosensitivity in *Drosophila*. *Nature* 454, 1014–1018.

Gould, J. L. (1975). Honey bee recruitment: The dance-language controversy. *Science* 189, 685–692.

Gould, J. L. (1980). The case for magnetic sensitivity in birds and bees. *American Scientist* 68, 256–267.

Gould, J. L. (1984). Processing of sun-azimuth information by honey bees. *Animal Behaviour* 32, 149–152.

Gould, J. L. (1998). Sensory bases of navigation. *Current Biology* 8, R731–R738.

Gould, J. L. (2010). Magnetoreception. *Current Biology* 20, R431–R435.

Gould, J. L., and Gould, C. G. (1995). *The Honey Bee*, 2nd ed. (New York, NY: W. H. Freeman).

Gould, J. L., Kirschvink, J. L., and Deffeyes, K. S. (1978). Bees have magnetic remanence. *Science* 201, 1026–1028.

Hsu, C.-Y., and Li, C.-W. (1994). Magnetoreception in honey bees. *Science* 265, 95–97.

Kirschvink, J. L., and Gould, J. L. (1981). Biogenic magnetite as a basis for magnetic field detection in animals. *BioSystems* 13, 181–201.

Leask, M. J. M. (1977). A physiochemical mechanism for magnetic-field detection. *Nature* 267, 144–145.

Lindauer, M., and Martin, H. (1968). Die Schwereorientierung der Bienen unter dem Einfluss der Erdmagnetfelds. *Journal of Comparative Physiology* 60, 219–243.

Lindauer, M., and Martin, H. (1972). Magnetic effects on dancing bees. In S. R. Galler, K. Schmidt-Koenig, G. J. Jacobs, and R. E. Belleville (Eds.), *Animal Orientation and Navigation* (Washington, DC: US Government Printing Office), pp. 559–567.

Lohmann, K. J. (1984). Magnetic remanence in the western Atlantic spiny lobster. *Journal of Experimental Biology* 113, 29–41.

Lohmann, K. J., Pentcheff, N. D., Nevitt, G. A., et al. (1995). Magnetic orientation of spiny lobsters in the ocean: Experiments with undersea coil systems. *Journal of Experimental Biology* 198, 2041–2048.

Lohmann, K. J., and Willows, A. O. D. (1987). Lunar-modulated geomagnetic orientation by a marine mollusk. *Science* 235, 3311–3334.

Lohmann, K. J., Willows, A. O. D., and Pinter, R. B. (1991). An identifiable molluscan neuron responds to changes in earth-strength magnetic fields. *Journal of Experimental Biology* 161, 1–24.

Martin, H., and Lindauer, M. (1977). Der Einfluss der Erdmagnetfelds und die Schwereorientierung der Honigbienen. *Journal of Comparative Physiology* 122, 145–187.

Merlin, C., Gegear, R. J., and Reppert, S. M. (2009). Antennal circadian clocks coordinate sun compass orientation in migratory monarch butterflies. *Science* 325, 1700–1704.

Papi, F., and Pardi, L. (1963). On the lunar orientation of sandhoppers. *Biological Bulletin* 124, 97–105.

Perez, S. M., Taylor, O. R., and Jander, R. (1997). Sun compass in monarch butterflies. *Nature* 387, 29.

Phillips, J. B., and Sayeed, O. (1993). Wavelength-dependent effects of light on magnetic compass orientation in *Drosophila melanogaster. Journal of Comparative Physiology A: Neuroethology, Sensory, Neural, and Behavioral Physiology* 172, 303–308.

Popescu, I. R., and Willows, A. O. D. (1999). Sources of magnetic sensory input to identified neurons active during crawling in the marine mollusk *Tritonia diomedea. Journal of Experimental Biology* 202, 3029–3036.

Quinn, T. P. (1980). Evidence for celestial and magnetic compass orientation in lake-migrating sockeye salmon fry. *Journal of Comparative Physiology* 137, 243–248.

Reid, S. F., Narendra, A., Hemmi, J. M., and Zeil, J. (2011). Polarised skylight and the landmark panorama provide night-active bull ants with compass information during route following. *Journal of Experimental Biology* 214, 363–370.

Renner, M. (1957). Neue Versuche über den Zeitsinn der Honigbienen. *Zeitschrift für Vergleichende Physiologie* 40, 85–118.

Reynolds, A. M, Reynolds, D. R., Smith, A. D., and Chapman, J. W. (2010). A single wind-mediated mechanism explains high-altitude "non-goal oriented" headings and layering of nocturnally migrating insects. *Proceedings of the Royal Society of London B: Biological Sciences* 277, 765–772.

Schmitt, D. E., and Esch, H. E. (1992). Magnetic orientation of honey bees in the laboratory. *Naturwissenschaften* 80, 41–43.

Scholz, A. T., Horrall, R. M., Cooper, J. C., and Hasler, A. D. (1976). Imprinting to chemical cues: The basis for home-stream selection in salmon. *Science* 192, 1247–1249.

Stephens, C. (2006). Bacterial cell biology: Managing magnetosomes. *Current Biology* 16, R363–R365.

Taylor, P. B. (1986). Experimental evidence for geomagnetic orientation in juvenile salmon. *Journal of Fish Biology* 28, 607–624.

Towne, W. F. (2008). Honeybees can learn the relationship between the solar ephemeris and a newly-experienced landscape. *Journal of Experimental Biology* 211, 3737–3743.

Towne, W. F., Baer, C. M., Fabiny, S. J., and Shinn, L. M. (2005). Does swarming cause bees to update their solar ephemeredes? *Journal of Experimental Biology* 208, 4049–4061.

Towne, W. F., and Moscrip, H. (2008). The connection between landscapes and the solar ephemeris in honey bees. *Journal of Experimental Biology* 211, 3729–3736.

Ugolini, A., Castellini, C., and Telli, L. C. (2007). Moon orientation on moonless nights. *Animal Behaviour* 73, 453–456.

Ugolini, A., Melis, C., Innocenti, R., et al. (1999). Moon and sun compasses in sandhoppers rely on two separate chronometric mechanisms. *Proceedings of the Royal Society of London B: Biological Sciences* 266, 749–752.

Ugolini, A., Morabito, F., and Taiti, S. (1995). Innate landward orientation in the littoral isopod *Tylos europaeus*. *Ethology Ecology & Evolution* 7, 387–391.

Ugolini, A., and Pezzani, A. (1995). Magnetic compass and learning of the Y-axis (sea–land) direction in the marine isopod *Idotea baltica basteri*. *Animal Behaviour* 50, 295–300.

von Frisch, K. (1967). *The Dance Language and Orientation of Bees* (Cambridge, MA: Harvard University Press).

Weaver, J. C., Vaughan, T. E., and Astumian, R. D. (2000). Biological sensing of small field differences by magnetically sensitive chemical reactions. *Nature* 405, 707–709.

Wehner, R., and Müller, M. (2006). The significance of direct sunlight and polarized skylight in the ant's celestial system of navigation. *Proceedings of the National Academy of Sciences* 103, 12575–12579.

Wiltschko, R., and Wiltschko, W. (1995). *Magnetic Orientation in Animals* (Berlin, Germany: Springer-Verlag).

Chapter 5

Able, K. P. (1989). Skylight polarization patterns and the orientation of migratory birds. *Journal of Experimental Biology* 141, 241–256.

Able, K. P., and Able, M. A. (1990). Calibration of the magnetic compass of a migratory bird by celestial rotation. *Nature* 347, 378–389.

Able, K. P., and Able, M. A. (1990). Ontogeny of migratory orientation in the savannah sparrow: Calibration of the magnetic compass. *Animal Behaviour* 39, 905–913.

Able, K. P., and Able, M. A. (1990). Ontogeny of migratory orientation in the savannah sparrow: Mechanisms at sunset. *Animal Behaviour* 39, 1189–1198.

Able, K. P., and Able, M. A. (1995). Interactions in the flexible orientation system of a migratory bird. *Nature* 375, 230–232.

Able, K. P., and Able, M. A. (1995). Manipulations of polarized skylight calibrate magnetic orientation in a migratory bird. *Journal of Comparative Physiology A: Neuroethology, Sensory, Neural, and Behavioral Physiology* 177, 351–356.

Able, K. P., and Able, M. A. (1996). The flexible migratory orientation system of the savannah sparrow. *Journal of Experimental Biology* 199, 3–8.

Åkesson, S., Morin, J., Muheim, R., and Ottosson, U. (2001). Avian orientation at steep angles of inclination: Experiments with migratory white-crowned sparrows at the magnetic north pole. *Proceedings of the Royal Society of London B: Biological Sciences* 268, 1907–1913.

Åkesson, S., Morin, J., Muheim, R., and Ottosson, U. (2002). Avian orientation: Effects of cue-conflict experiments with young migratory songbirds in the high Arctic. *Animal Behaviour* 64, 469–475.

Beason, R. C., and Semm, P. (1987). Magnetic responses of the trigeminal nerve system of the bobolink. *Neuroscience Letters* 80, 229–234.

Beason, R. C., Wiltschko, R., and Wiltschko, W. (1997). Pigeon homing: Effects of magnetic pulses on initial orientation. *Auk* 114, 405–415.

Begall, S., Cerveny, J., Neef, J., et al. (2008). Magnetic alignment in grazing and resting cattle and deer. *Proceedings of the National Academy of Sciences* 105, 13451–13455.

Bletz, H., Weindler, P., Wiltschko, R., and Wiltschko, W. (1996). The magnetic field as reference for the innate migratory direction in blackcaps. *Naturwissenschaften* 83, 430–432.

Davila, A. F., Winkelhofer, M., Shcherbakov, V. P., and Petersen, N. (2005). Magnetic pulse affects a putative magnetoreceptor mechanism. *Biophysical Journal* 89, 56–63.

Deutschlander, M. E., Borland, S. C., and Phillips, J. B. (1999). Extraocular magnetic compass in newts. *Nature* 400, 324–325.

Diebel, C. E., Proksch, R., Green, C. R., et al. (2000). Magnetite defines a magnetoreceptor. *Nature* 406, 299–302.

Eimas, P. D., Siqueland, E. R., Jusczyk, P., and Vigorito, J. (1971). Speech perception in infants. *Science* 171, 303–306.

Elsner, B. (1978). Accurate measurements of the initial tracks of homing pigeons. In K. Schmidt-Koenig and W. T. Keeton (Eds.), pp. 194–198.

Emlen, S. T. (1967). Migratory orientation in the indigo bunting. Part I: Evidence for the use of celestial cues. *Auk* 84, 309–342.

Emlen, S. T. (1967). Migratory orientation in the indigo bunting. Part II: Mechanisms of celestial orientation. *Auk* 84, 463–489.

Emlen, S. T. (1970). Celestial rotation: Its importance in the development of migratory orientation. *Science* 170, 1198–1201.

Freire, R., Munro, U. H., Rogers, L. J., et al. (2005). Chickens orient using a magnetic compass. *Current Biology* 15, R620–R621.

Gould, J. L. (1980). The case for magnetic sensitivity in birds and bees. *American Scientist* 68, 256–267.

Gould, J. L. (1995). Constant compass calibration. *Nature* 375, 184.

Gould, J. L. (1995). Fly (almost) south young bird. *Nature* 383, 123–124.

Gould, J. L. (1998). Sensory bases of navigation. *Current Biology* 8, R731–R738.

Gould, J. L. (2008). Animal navigation: Evolution of magnetic orientation. *Current Biology* 18, R482–R484.

Gould, J. L. (2010). Magnetoreception. *Current Biology* 20, R431–R435.

Gould, J. L., and Marler, P. (1987). Learning by instinct. *Scientific American* 256 (1), 74–85.

Griffin, D. R. (1973). Oriented bird migration in or between opaque cloud layers. *Proceedings of the American Philosophical Society* 117, 117–141.

Griffin, D. R. (1974). Sounds audible to migrating birds. *Animal Behaviour* 22, 672–678.

Griffin, D. R. (1976). The audibility of frog choruses to migrating birds. *Animal Behaviour* 24, 421–427.

Haugh, C. V., and Walker, M. M. (1998). Magnetic discrimination learning in rainbow trout. *Journal of Navigation* 51, 35–45.

Hawryshyn, C. W. (1992). Polarization vision in fish. *American Scientist* 80, 164–175.

Holland, R. A., Thorup, K., Vonhof, M. J., et al. (2006). Bat orientation using the earth's magnetic field. *Nature* 444, 702.

Kalmijn, A. (1978). Experimental evidence of geomagnetic orientation in elasmobranch fishes. In K. Schmidt-Koenig and W. T. Keeton (Eds.), pp. 345–353.

Keeton, W. T. (1969). Orientation by pigeons: Is the sun necessary? *Science* 165, 922–928.

Keeton, W. T. (1971). Magnets interfere with pigeon homing. *Proceedings of the National Academy of Sciences* 68, 102–106.

Keeton, W. T. (1974). The mystery of pigeon homing. *Scientific American* 231 (6), 96–104.

Kimchi, T., Etienne, A. S., and Terkel, J. (2004). A subterranean mammal uses the magnetic compass for path integration. *Proceedings of the National Academy of Sciences* 101, 1105–1109.

Kirschvink, J. L. (1997). Homing in on vertebrates. *Nature* 390, 339–340.

Kirschvink, J. L., and Gould, J. L. (1981). Biogenic magnetite as a basis for magnetic field detection in animals. *BioSystems* 13, 181–201.

Kreithen, M. L., and Keeton, W. T. (1974). Detection of polarized light by the homing pigeon. *Journal of Comparative Physiology* 89, 83–92.

Larkin, T., and Keeton, W. T. (1978). An apparent lunar rhythm in day-to-day variations in initial bearings of homing pigeons. In K. Schmidt-Koenig and W. T. Keeton (Eds.), pp. 92–106.

Leask, M. J. M. (1977). A physiochemical mechanism for magnetic-field detection. *Nature* 267, 144–145.

Lohmann, K. J. (1991). Magnetic orientation by hatchling loggerhead sea turtles. *Journal of Experimental Biology* 155, 37–49.

Lohmann, K. J. (2007). Sea turtles: Navigating with magnetism. *Current Biology* 17, R102–R104.

Lohmann, K. J. (2010). Magnetic-field perception. *Nature* 464, 1140–1142.

Lohmann, K. J., and Johnsen, S. (2000). Neurobiology of magnetoreception in vertebrate animals. *Trends in Neurosciences* 23, 153–159.

Lohmann, K. J., and Lohmann, C. M. F. (1992). Orientation to oceanic waves by green turtle hatchlings. *Journal of Experimental Biology* 171, 1–13.

Lohmann, K. J., and Lohmann, C. M. F. (1993). A light-dependent magnetic compass in the leatherback sea turtle. *Biological Bulletin* 185, 149–151.

Lohmann, K. J., and Lohmann, C. M. F. (1994). Acquisition of magnetic directional preference in hatchling loggerhead sea turtles. *Journal of Experimental Biology* 190, 1–8.

Lohmann, K. J., and Lohmann, C. M. F. (1996). Orientation and open-sea navigation in sea turtles. *Journal of Experimental Biology* 199, 73–81.

Lohmann, K. J., Swartz, A. W., and Lohmann, C. M. F. (1995). Perception of ocean wave direction by sea turtles. *Journal of Experimental Biology* 198, 1079–1085.

Maeda, K., Henbest, K. B., Cintolesi, F., et al. (2008). Chemical compass model of avian magnetoreception. *Nature* 453, 387–390.

Marhold, S., Wiltschko, W., and Burda, H. (1997). A magnetic polarity compass for direction finding in a subterranean mammal. *Naturwissenschaften* 84, 421–423.

Michener, M., and Walcott, C. (1967). Homing of single pigeons—Analysis of tracks. *Journal of Experimental Biology* 47, 99–131.

Moore, F. R. (1978). Sunset and the orientation of a nocturnal migrant bird. *Nature* 274, 154–156.

Mora, C. V., Davison, M., Wild, J. M., and Walker, M. M. (2004). Magnetoreception and its trigeminal mediation in the homing pigeons. *Nature* 432, 508–511.

Mouritsen, H., Feenders, G., Leidvogel, M., and Kropp, W. (2004). Migratory birds use head scans to detect the direction of the earth's magnetic field. *Current Biology* 14, 1946–1949.

Mouritsen, H., Janssen-Bienhold, U., Liedvogel, M., et al. (2004). Cryptochromes and neuronal-activity markers colocalize in the retina of migratory birds during magnetic orientation. *Proceedings of the National Academy of Sciences* 101, 14294–14299.

Muheim, R., and Åkesson, S. (2002). Clock shift experiments with savannah sparrows at high northern latitudes. *Behavioral Ecology and Sociobiology* 51, 394–401.

Muheim, R., Phillips, J. B., and Åkesson, S. (2006). Polarized light cues underlie compass calibration in migratory songbirds. *Science* 313, 837–839.

Munro, U., Munro, J., Phillips, J. B., and Wiltschko, W. (1997). Effect of wavelength of light and pulse magnetization on different magnetoreception systems in a migrating bird. *Australian Journal of Zoology* 45, 189–198.

Munro, U., and Wiltschko, R. (1993). Magnetic compass orientation in the yellow-faced honeyeater. *Behavioral Ecology and Sociobiology* 32, 141–145.

Munro, U., and Wiltschko, R. (1995). The role of skylight polarization on the orientation of a day-migrating bird species. *Journal of Comparative Physiology* 177, 357–362.

Murphey, P. A. (1981). Celestial compass orientation in juvenile alligators. *Copia* 1981, 638–645.

Nemec, P., Altmann, J., Marhold, S., et al. (2001). Neuroanatomy of magnetoreception: The superior colliculus involved in magnetic orientation in a mammal. *Science* 294, 366–368.

Papi, F., and Wallraff, H. G. (1982). *Avian Navigation* (Berlin, Germany: Springer-Verlag).

Phillips, J. B. (1986). Two magnetoreception pathways in a migratory salamander. *Science* 233, 765–767.

Phillips, J. B., and Borland, S. C. (1992). Behavioural evidence for use of a light-dependent magnetoreception mechanism by a vertebrate. *Nature* 359, 142–144.

Phillips, J. B., Borland, S. C., Freake, M. J., et al. (2002). "Fixed-axis" magnetic orientation by an amphibian. *Journal of Experimental Biology* 205, 3903–3914.

Phillips, J. B., Muheim, R., and Jorge, P. E. (2010). A behavioral perspective on the biophysics and light-dependent magnetic compass: A link between directional and spatial perception? *Journal of Experimental Biology* 213, 3247–3255.

Prinz, K., and Wiltschko, W. (1992). Migratory orientation of pied flycatchers: Interaction of stellar and magnetic information during ontogeny. *Animal Behaviour* 44, 539–545.

Quine, D. B., and Kreithen, M. L. (1981). Frequency shift discrimination: Can homing pigeons locate infrasounds by Doppler shifts? *Journal of Comparative Physiology* 141, 153–155.

Ritz, T., Adem, S., and Schulten, K. (2000). A model for photoreceptor-based magnetoreception in birds. *Biophysical Journal* 78, 707–718.

Ritz, T., Phillips, J. B., and Dommer, D. H. (2002). Shedding light on vertebrate magnetoreception. *Neuron* 34, 503–506.

Ritz, T., Thalau, P., Phillips, J. B., et al. (2004). Avian magnetic compass: Resonance effects indicate a radical pair mechanism. *Nature* 429, 177–180.

Sandberg, R. (1991). Sunset orientation of robins with different fields of sky vision. *Behavioral Ecology and Sociobiology* 28, 77–83.

Sandberg, R., Bäckman, J., Moore, F. R., and Lõhmus, M. (2000). Magnetic information calibrates celestial cues during migration. *Animal Behaviour* 60, 453–463.

Schmidt-Koenig, K. (1975). *Avian Orientation and Navigation* (New York, NY: Academic Press).

Schmidt-Koenig, K., and Keeton, W. T. (Eds.). (1978). *Animal Migration, Navigation, and Homing* (Berlin, Germany: Springer-Verlag).

Schmidt-Koenig, K., and Walcott, C. (1978). Tracks of pigeons homing with frosted lenses. *Animal Behaviour* 26, 480–486.

Stapput, K., Güntürkün, O., Hoffmann, K.-P., et al. (2010). Magnetoreception of directional information in birds requires nondegraded vision. *Current Biology* 18, 602–606.

Stapput, K., Thalau, P., Wiltschko, R., and Wiltschko, W. (2008). Orientation of birds in total darkness. *Current Biology* 20, 1259–1262.

Stutchbury, B. J. M., Tarof, S. A., Done, T., et al. (2009). Tracking long-distance songbird migration by using geolocators. *Science* 323, 896.

Trut, L. M. (1999). Early canid domestication: The farm-fox experiment. *American Scientist* 87, 160–169.

Walcott, C., Gould, J. L., and Kirschvink, J. L. (1979). Pigeons have magnets. *Science* 205, 1027–1029.

Walcott, C., and Green, R. P. (1974). Orientation of homing pigeons altered by a change in the direction of an applied magnetic field. *Science* 184, 180–182.

Walcott, C., Lednor, A. J., and Gould, J. L. (1988). Homing of pigeons after treatment with strong magnetic fields. *Journal of Experimental Biology* 134, 27–41.

Walcott, C., and Schmidt-Koenig, K. (1973). The effect on pigeon homing of anesthesia during displacement. *Auk* 90, 281–286.

Walker, M. M. (1984). Learned magnetic field discrimination in yellowfin tuna. *Journal of Comparative Physiology A: Neuroethology, Sensory, Neural, and Behavioral Physiology* 155, 673–679.

Walker, M. M., Diebel, C. E., Haugh, C. V., et al. (1997). Structure and function of the vertebrate magnetic sense. *Nature* 390, 371–376.

Weaver, J. C., Vaughan, T. E., and Astumian, R. D. (2000). Biological sensing of small field differences by magnetically sensitive chemical reactions. *Nature* 405, 707–709.

Weindler, P., Wiltschko, R., and Wiltschko, W. (1996). Magnetic information affects the stellar orientation of young bird migrants. *Nature* 383, 158–160.

Williams, M. N., and Wild, J. M. (2001). Trigeminally innervated iron-containing structures in the beak of homing pigeons, and other birds. *Brain Research* 899, 243–246.

Wiltschko, R. (1991). Role of experience in avian navigation and homing. In P. Berthold (Ed.), *Orientation in Birds* (Basel, Switzerland: Birkhäuser Verlag), pp. 250–269.

Wiltschko, R., Kumpfmüller, R., Muth, R., and Wiltschko, W. (1994). Pigeon homing: The effect of a clock shift is often smaller than predicted. *Behavioral Ecology and Sociobiology* 35, 63–73.

Wiltschko, R., Nohr, D., and Wiltschko, W. (1981). Pigeons with a deficient sun compass use the magnetic compass. *Science* 214, 343–345.

Wiltschko, R., Walker, M., and Wiltschko, W. (2000). Sun-compass orientation in homing pigeons: Compensation for different rates of change in azimuth? *Journal of Experimental Biology* 203, 889–894.

Wiltschko, R., and Wiltschko, W. (1972). Magnetic compass of European robins. *Science* 176, 62–64.

Wiltschko, R., and Wiltschko, W. (1975). The interaction of stars and magnetic fields in the orientation system of night-migrating birds. *Zeitschift für Tierpsychologie* 35, 536–542.

Wiltschko, R., and Wiltschko, W. (1985). Pigeon homing: Change in navigational strategy during ontogeny. *Animal Behaviour* 33, 583–590.

Wiltschko, R., and Wiltschko, W. (1995). *Magnetic Orientation in Animals* (Berlin, Germany: Springer-Verlag).

Wiltschko, R., and Wiltschko, W. (1996). Magnetic orientation in birds. *Journal of Experimental Biology* 199, 29–38.

Wiltschko, R., and Wiltschko, W. (1998). Pigeon homing: Effect of various wavelengths of light during displacement. *Naturwissenschaften* 85, 164–167.

Wiltschko, R., and Wiltschko, W. (1998). The navigation system of birds. In R. P. Balda, I. M. Pepperberg, and A. C. Kamil (Eds.), *Animal Cognition in Nature* (San Diego, CA: Academic Press), pp. 155–199.

Wiltschko, R., and Wiltschko, W. (2000). Light-dependent magnetoreception in birds: Does directional information change with light intensity? *Naturwissenschaften* 87, 36–40.

Wiltschko, R., and Wiltschko, W. (2001). Clock shift experiments with homing pigeons: A compromise between solar and magnetic information? *Behavioral Ecology and Sociobiology* 49, 393–400.

Wiltschko, W., Munro, U., Beason, R. C., et al. (1994). A magnetic pulse leads to a temporary deflection in the orientation of migratory birds. *Experientia* 50, 697–700.

Wiltschko, W., Munro, U., Ford, H., and Wiltschko, R. (1993). Red light disrupts magnetic orientation of migratory birds. *Nature* 364, 525–527.

Wiltschko, W., Munro, U., Ford, H., and Wiltschko, R. (1999). After-effects of exposure to conflicting celestial and magnetic cues at sunset in migratory silvereyes. *Journal of Avian Biology* 30, 56–62.

Wiltschko, W., Munro, U., and Wiltschko, R. (1997). Magnetoreception in migratory birds: Light-mediated and magnetite-mediated processes? In *Orientation & Navigation—Birds, Humans, and Other Animals* (Oxford, United Kingdom: Royal Institute of Navigation), pp. 1.1–1.9.

Wiltschko, W., Traudt, J., Güntürkün, O., et al. (2002). Lateralization of magnetic compass orientation in a migratory bird. *Nature* 419, 467–470.

Wiltschko, W., Weindler, P., and Wiltschko, R. (1998). Interaction of magnetic and celestial cues in the migratory orientation of passerines. *Journal of Avian Biology* 29, 606–617.

Wiltschko, W., and Wiltschko, R. (2002). Magnetic compass orientation in birds and its physiological basis. *Naturwissenschaften* 89, 445–452.

Wiltschko, W., Wiltschko, R., and Munro, U. (1997). Migratory orientation in birds: The effects and after-effects of exposure to conflicting celestial and magnetic cues. In *Orientation & Navigation—Birds, Humans, and Other Animals* (Oxford, United Kingdom: Royal Institute of Navigation), pp. 6.1–6.14.

Yodlowski, M. L., Kreithen, M. L., and Keeton, W. T. (1977). Detection of atmospheric infrasound by homing pigeons. *Nature* 265, 725–726.

Zapka, M., Heyers, D., Hein, C. M., et al. (2009). Visual but not trigeminal mediation of magnetic compass information in a migratory bird. *Nature* 461, 1274–1277.

Chapter 6

Balda, R. P., and Wiltschko, W. (1991). Caching and recovery in scrub jays. *Condor* 93, 1020–1023.

Biro, D., Meade, J., and Guilford, T. (2004). Familiar route loyalty implies visual pilotage in the homing pigeon. *Proceedings of the National Academy of Sciences* 101, 17440–17443.

Bühlmann, C., Cheng, K., and Wehner, R. (2011). Vector-based and landmark-guided navigation in desert ants inhabiting landmark-free and landmark-rich environments. *Journal of Experimental Biology* 214, 2845–2853.

Chapuis, N., and Scardigli, P. (1993). Shortcut ability in hamsters. *Animal Learning and Behavior* 21, 255–265.

Cheng, K. (2000). How honey bees find a place: Lessons from a simple mind. *Animal Learning and Behavior* 28, 1–15.

Collett, M. (2009). Spatial memories in insects. *Current Biology* 19, R1103–R1108.

Collett, M., and Collett, T. S. (2009). Learning and maintenance of local vectors in desert ant navigation. *Journal of Experimental Biology* 212, 895–900.

Collett, M., and Collett, T. S. (2009). Local and global navigational coordinate systems in desert ants. *Journal of Experimental Biology* 212, 901–905.

Collett, M., Collett, T. S., Bisch, S., and Wehner, R. (1998). Local and global vectors in desert ant navigation. *Nature* 394, 269–272.

Collett, T. S. (1996). Insect navigation *en route* to the goal: Multiple strategies for the use of landmarks. *Journal of Experimental Biology* 199, 227–235.

Cook, R. G., Brown, M. F., and Riley, D. A. (1983). Flexible memory processing by rats: Use of prospective and retrospective information in the radial maze. *Journal of Experimental Psychology: Animal Behavior Processes* 11, 453–469.

Dyer, F. C. (1991). Bees acquire route-based memories but not cognitive maps in a familiar landscape. *Journal of Comparative Physiology A: Neuroethology, Sensory, Neural, and Behavioral Physiology* 160, 621–633.

Etienne, A. S., Maurer, J., Berlie, J., et al. (1998). Navigation through vector addition. *Nature* 396, 161–164.

Etienne, A. S., Maurer, J., and Séguinot, V. (1996). Path integration in mammals and its interaction with visual landmarks. *Journal of Experimental Biology* 199, 201–209.

Galea, L. A. M., Kavaliers, M., and Ossenkopp, K.-P. (1996). Sexually dimorphic spatial learning in meadow voles and deer mice. *Journal of Experimental Biology* 199, 195–200.

Gallistel, C. R., and Cramer, A. E. (1996). Computations on metric maps in mammals: Getting oriented and choosing a multi-destination route. *Journal of Experimental Biology* 199, 211–217.

Gould, J. L. (1986). The locale map of honey bees: Do insects have cognitive maps? *Science* 232, 861–863.

Gould, J. L. (1986). Pattern learning in honey bees. *Animal Behaviour* 34, 990–997.

Gould, J. L. (1987). Landmark learning in honey bees. *Animal Behaviour* 35, 26–34.

Gould, J. L., and Gould, C. G. (1995). *The Honey Bee*, rev. ed. (New York, NY: W. H. Freeman).

Hermer, L., and Spelke, E. (1994). A geometric process for spatial reorientation in young children. *Nature* 370, 57–59.

Hermer, L., and Spelke, E. (1996). Modularity and development: The case of spatial reorientation. *Cognition* 61, 195–232.

Isack, H. A., and Reyer, H. U. (1989). Honeyguides and honey gathers. *Science* 243, 1343–1346.

Jacobs, L. F., and Schenk, F. (2003). Unpacking the cognitive map: The parallel map theory of hippocampal function. *Psychological Review* 110, 285–315.

Judd, S. P. D., and Collett, T. S. (1998). Multiple stored views and landmark guidance in ants. *Nature* 392, 710–714.

Kamil, A. C., and Jones, J. E. (1997). The seed-storing corvid Clark's nutcracker learns geometric relationships between landmarks. *Nature* 390, 276–278.

Kavanau, J. L. (1969). Behaviour of captive white-footed mice. In E. R. Willems and H. L. Raush (Eds.), *Naturalistic Viewpoints in Psychology* (New York, NY: Holt, Rinehart, Winston), pp. 221–270.

Kimchi, T., Etienne, A. S., and Terkel, J. (2004). A subterranean mammal uses the magnetic compass for path integration. *Proceedings of the National Academy of Sciences* 101, 1105–1109.

Köhler, W. (1927). *The Mentality of Apes* (New York, NY: Harcourt Brace).

Langston, R. F., Ainge, J. A., Couey, J. J., et al. (2010). Development of the spatial representation system in the rat. *Science* 328, 1576–1580.

Lehrer, M. (1996). Small-scale navigation in the honey bee: Active acquisition of visual information about the goal. *Journal of Experimental Biology* 199, 253–261.

Lipp, H.-P., Vyssotski, A., and Wolfer, D. P. (2004). Pigeon homing along highways and exits. *Current Biology* 14, 1239–1249.

Maguire, E. A., Burgess, N., Donnett, J. G., et al. (1998). Knowing where and getting there: A human navigation network. *Science* 280, 921–924.

McNaughton, B. L., Barnes, C. A., Gerrard, J. L., et al. (1996). Deciphering the hippocampal polyglot: The hippocampus as a path-integration system. *Journal of Experimental Biology* 199, 173–185.

Menzel, E. W. (1973). Chimpanzee spatial memory organization. *Science* 182, 943–945.

Menzel, R., Chittka, L., Eichmüller, S., et al. (1990). Dominance of celestial cues over landmarks disproves map-like orientation in honey bees. *Zeitschrift für Naturforschung C: A Journal of Biosciences* 45, 723–726.

Menzel, R., Greggers, U., Smith, A., et al. (2005). Honey bees navigate according to a map-like spatial memory. *Proceedings of the National Academy of Sciences* 102, 3040–3045.

Menzel, R., Kirbach, A., Haass, W.-D., et al. (2011). A common frame of reference for learned and communicated vectors in honey bee navigation. *Current Biology* 21, 645–650.

Michener, M., and Walcott, C. (1967). Homing of single pigeons—Analysis of tracks. *Journal of Experimental Biology* 47, 99–131.

Nardini, M., Jones, P., Bedford, R., and Braddick, O. (2008). Development of cue integration in human navigation. *Current Biology* 18, 689–693.

Normand, E., and Boesch, C. (2009). Sophisticated Euclidean maps in forest chimpanzees. *Animal Behaviour* 77, 1195–1201.

Olton, D. S. (1977). Spatial memory. *Scientific American* 236 (6), 82–98.

Olton, D. S. (1978). Characteristics of spatial memory. In S. H. Hulse, H. Fowler, and W. K. Honig (Eds.), *Cognitive Processes in Animal Behavior* (Hillsdale, NJ: Erlbaum), pp. 341–373.

Pearce, J. M., Roberts, A. D. L., and Good, M. (1998). Hippocampal lesions disrupt navigation based on cognitive maps but not heading vectors. *Nature* 396, 75–77.

Reznikova, Z. (2007). *Animal Intelligence* (Cambridge, United Kingdom: Cambridge University Press).

Schiller, P. H. (1957). Innate motor action as a basis of learning. In C. H. Schiller (Ed.), *Instinctive Behavior* (New York, NY: International Universities Press), pp. 264–287.

Souman, J. L., Frissen, I., Sreenlvasa, M. N., and Ernst, M. O. (2009). Walking straight into circles. *Current Biology* 19, 1538–1542.

Tarsitano, M. S., and Jackson, R. R. (1994). Jumping spiders make predatory detours requiring movement away from prey. *Behaviour* 131, 65–73.

Tinbergen, N., and Kruyt, W. (1938). Über die Orientierung des Bienenwolfes III: Die Bevorzugung bestimmter Wegmarken. *Zeitschrift für Vergleichende Physiologie* 25, 292–234.

Tolman, E. C. (1948). Cognitive maps in rats and men. *Psychological Review* 55, 189–208.

Tomback, D. F. (1980). How nutcrackers find their seed stores. *Condor* 82, 10–19.

von Frisch, K. (1967). *The Dance Language and Orientation of Bees* (Cambridge, MA: Harvard University Press).

Walcott, C., and Schmidt-Koenig, K. (1973). The effect on homing of anesthesia during displacement. *Auk* 90, 281–286.

Watson, J. B. (1924). *Behaviorism* (New York, NY: The People's Institute).

Wehner, R., and Müller, M. (2010). Piloting in desert ants: Pinpointing the goal by discrete landmarks. *Journal of Experimental Biology* 213, 4174–4179.

Wills, T. J., Cacucci, F., Burgess, N., and O'Keefe, J. (2010). Development of the hippocampal cognitive map in preweanling rats. *Science* 328, 1573–1576.

Wiltschko, R. (1991). The role of experience in avian navigation and homing. In P. Berthold (Ed.), *Orientation in Birds* (Basel, Switzerland: Birkhäuser Verlag), pp. 250–269.

Wiltschko, W., and Balda, R. P. (1989). Sun-compass orientation in seed-caching scrub jays. *Journal of Comparative Physiology* 164, 717–721.

Wittlinger, M., Wehner, R., and Wolf, H. (2006). The ant odometer: Stepping on stilts and stumps. *Science* 312, 1965–1967.

Wynne, C. D. L. (2001). *Animal Cognition* (New York, NY: Palgrave).

Chapter 7

Able, K. P. (1995). Orientation and navigation: A perspective on fifty years of research. *Condor* 97, 592–604.

Able, K. P. (1996). The debate over olfactory navigation by homing pigeons. *Journal of Experimental Biology* 199, 121–124.

Åkesson, S., Ottosson, U., and Sandberg, R. (1995). Bird migration: Displacement experiments with young autumn migrating wheatears along the Arctic coast of Russia. *Proceedings of the Royal Society of London B: Biological Sciences* 262, 189–195.

Alerstam, T. (2003). The lobster navigators. *Nature* 421, 27–28.

Alerstam, T. (2006). Conflicting evidence about long-distance animal navigation. *Science* 313, 791–794.

Baker, R. R. (1980). Goal orientation by blindfolded humans after long-distance displacement: Possible involvement of a magnetic sense. *Science* 210, 555–557.

Baker, R. R. (1981). *Human Navigation and the Sixth Sense* (London, United Kingdom: Hodder and Stoughton).

Baker, R. R. (1984). *Bird Navigation: The Solution of a Mystery?* (London, United Kingdom: Hodder and Stoughton).

Beason, R. C., and Semm, P. (1987). Magnetic responses of the trigeminal nerve system of the bobolink. *Neuroscience Letters* 80, 229–234.

Beason, R. C., Wiltschko, R., and Wiltschko, W. (1997). Pigeon homing: Effects of magnetic pulses on initial orientation. *Auk* 114, 405–415.

Benvenuti, S., Baldaccini, N. E., and Ioalè, P. (1982). Pigeon homing: Effect of altered magnetic field during displacement on initial orientation. In F. Papi and H. G. Wallraff (Eds.), pp. 140–148.

Benvenuti, S., Bingman, V. P., and Gagliardo, A. (1998). Effect of zinc-sulphate induced anosmia on pigeon homing: A comparison among birds in different regions. *Trends in Comparative Biochemistry and Physiology* 5, 221–228.

Benvenuti, S., Ioalè, P., and Nacci, L. (1994). A new experiment to verify the spatial range of pigeons' olfactory map. *Behaviour* 131, 277–292.

Berthold, P. (1993). *Bird Migration: A General Survey* (Oxford, United Kingdom: Oxford University Press).

Bingman, V. P., Alyan, S., and Benvenuti, S. (1998). The importance of atmospheric odors for the homing performance of pigeons in the Sonoran Desert of the southwestern United States. *Journal of Experimental Biology* 201, 755–760.

Bingman, V. P., and Benvenuti, S. (1996). Olfaction and the homing ability of pigeons in the southeastern United States. *Journal of Experimental Zoology* 60, 186–192.

Bingman, V. P., and Cheng, K. (2005). Mechanisms of animal global navigation: Comparative perspectives and enduring challenges. *Ethology Ecology & Evolution* 17, 295–318.

Boles, L. C., and Lohmann, K. J. (2003). True navigation and magnetic maps in spiny lobsters. *Nature* 421, 60–63.

Chernetsov, N., Kishkinev, D., and Mouritsen, H. (2008). A long-distance avian migrant compensates for longitudinal displacement during spring migration. *Current Biology* 18, 188–190.

Davila, A. F., Winklhofer, M., Shcherbakov, V. P., and Petersen, N. (2005). Magnetic pulse affects a putative magnetoreceptor mechanism. *Biophysical Journal* 89, 56–63.

Dennis, T. E., Rayner, M. J., and Walker, M. M. (2007). Orientation to geomagnetic intensity in homing pigeons. *Proceedings of the Royal Society of London B: Biological Sciences* 274, 1153–1158.

Dornfeldt, K. (1996). Pigeon homing in the meteorological and solar-geomagnetic environment: What pigeon race data say. *Ethology* 102, 413–435.

Etienne, A. S., Maurer, J., Berlie, J., et al. (1998). Navigation through vector addition. *Nature* 396, 161–164.

Fildes, B. N., O'Loughlin, B. J., Bradshaw, J. L., and Ewens, W. J. (1984). Human orientation with restricted sensory information: No evidence for magnetic sensitivity. *Perception* 13, 229–236.

Fischer, J. H., Freake, M. J., Borland, S. C., and Phillips, J. B. (2001). Evidence for the use of magnetic map information by an amphibian. *Animal Behaviour* 62, 1–10.

Foà, A., Wallraff, H. G., and Benvenuti, S. (1982). Comparative investigations of pigeon homing in Germany and Italy. In F. Papi and H. G. Wallraff (Eds.), pp. 232–238.

Fransson, T., Jakobsson, S., Johansson, P., et al. (2001). Magnetic cues trigger extensive refueling. *Nature* 414, 35.

Freake, M. J., Muheim, R., and Phillips, J. B. (2006). Magnetic maps in animals: A theory comes of age? *Quarterly Review of Biology* 81, 327–347.

Frei, U. (1982). Homing pigeons' behaviour in the irregular magnetic field of western Switzerland. In F. Papi and H. G. Wallraff (Eds.), pp. 129–139.

Frei, U., and Wagner, G. (1976). Die Anfangsorientierung von Brieftauben im erdmagnetisch gestörten Gebiet des Mont Jorat. *Revue Suisse de Zoologie* 83, 891–897.

Fuxjager, M. J., Eastwood, B. S., and Lohmann, K. J. (2011). Orientation of hatchling loggerhead sea turtles to regional magnetic fields along a transoceanic migratory pathway. *Journal of Experimental Biology* 214, 2504–2508.

Gagliardo, A., Filannino, C., Ioalè, P., et al. (2011). Olfactory lateralization in homing pigeons: A GPS study of birds released with unilateral olfactory inputs. *Journal of Experimental Biology* 214, 593–598.

Gagliardo, A., Ioalè, P., Savini, M., and Wild, J. M. (2006). Having the nerve to home: Trigeminal magnetoreceptor versus olfactory mediation of homing in pigeons. *Journal of Experimental Biology* 209, 2888–2892.

Gagliardo, A., Ioalè, P., Savini, M., and Wild, J. M. (2008). Navigational abilities of homing pigeons deprived of olfactory or trigeminally mediated magnetic information when young. *Journal of Experimental Biology* 211, 2046–2051.

Gill, R. E., Piersma, T., Hufford G., et al. (2005). Crossing the ultimate ecological barrier: Evidence for an 11,000-km–long nonstop flight from Alaska to New Zealand and eastern Australia by bar-tailed godwits. *Condor* 107, 1–20.

Gill, R. E., Tibbitts, T. L., Douglas, D. C., et al. (2009). Extreme endurance flights by landbirds crossing the Pacific Ocean: Ecological corridor rather than barrier? *Proceedings of the Royal Society B: Biological Sciences* 276, 447–457.

Gould, J. L. (1976). The dance-language controversy. *Quarterly Review of Biology* 51, 211–44.

Gould, J. L. (1980). The case for magnetic sensitivity in birds and bees. *American Scientist* 68, 256–267.

Gould, J. L. (1982). The map sense of pigeons. *Nature* 296, 205–211.

Gould, J. L. (1998). Sensory bases of navigation. *Current Biology* 8, R731–R738.

Gould, J. L. (2008). Animal navigation: The longitude problem. *Current Biology* 18, R214–R216.

Gould, J. L. (2008). Animal navigation: Evolution of magnetic orientation. *Current Biology* 18, R482–R484.

Gould, J. L. (2009). Animal navigation: A wake-up call for homing. *Current Biology* 19, R338–R339.

Gould, J. L. (2010). Magnetoreception. *Current Biology* 20, R431–R435.

Gould, J. L. (2011). Animal navigation: Longitude at last. *Current Biology* 21, R225–R227.

Gould, J. L., and Able, K. P. (1981). Human homing: An elusive phenomenon. *Science* 212, 1061–1063.

Graue, L. C. (1965). Initial orientation in pigeon homing related to magnetic contours. *American Zoologist* 5, 704.

Graue, L. C. (1970). Orientation and distance in pigeon homing. *Animal Behaviour* 18, 36–40.

Grönroos, J., Muheim, R., and Åkesson, S. (2010). Orientation and autumn migration routes of juvenile sharp-tailed sandpipers at a staging site in Alaska. *Journal of Experimental Biology* 213, 1829–1835.

Grüter, M., Wiltschko, R., and Wiltschko, W. (1982). Distribution of release-site biases around Frankfurt a.M., Germany. In F. Papi and H. G. Wallraff (Eds.), pp. 222–231.

Harada, Y. (2002). Experimental analysis of behavior of homing pigeons as a result of functional disorders of their lagena. *Acta Otolaryngologica* 122, 132–137.

Harada, Y., Taniguchi, M., Namatame, H., and Iida, A. (2001). Magnetic materials in otoliths of bird and fish lagena and their function. *Acta Otolaryngologica* 121, 590–595.

Heyers, D., Zapka, M., Hoffmeister, M., et al. (2010). Magnetic field changes activate the trigeminal brainstem complex in a migratory bird. *Proceedings of the National Academy of Sciences* 107, 9394–9399.

Holland, R. A., Thorup, K., Gagliardo, A., et al. (2009). Testing the role of sensory systems in the migratory heading of a songbird. *Journal of Experimental Biology* 212, 4065–4071.

Ioalè, P., Nozzolini, M., and Papi, F. (1990). Pigeons do extract directional information from olfactory stimuli. *Behavioral Ecology and Sociobiology* 26, 301–305.

Jorge, P. E., Marques, A. E., and Philips, J. B. (2009). Activational rather than navigational effects of odors on pigeon homing. *Current Biology* 19, 650–654.

Keeton, W. T. (1971). Magnets interfere with pigeon homing. *Proceedings of the National Academy of Sciences* 68, 102–106.

Keeton, W. T. (1973). Release site bias as a possible guide to the "map" component in pigeon homing. *Journal of Comparative Physiology* 86, 1–16.

Keeton, W. T. (1974). The orientational and navigational basis of homing in birds. *Advances in the Study of Behavior* 5, 47–132.

Keeton, W. T. (1974). The mystery of pigeon homing. *Scientific American* 231 (6), 96–104.

Keeton, W. T., Kreithen, M. L., and Hermayer, K. L. (1977). Orientation by pigeons deprived of olfaction by nasal tubes. *Journal of Comparative Physiology* 114, 289–299.

Keeton, W. T., Larkin, T. S., and Windsor, D. T. (1974). Normal fluctuations in the earth's magnetic field influence pigeon orientation. *Journal of Comparative Physiology* 95, 95–103.

Kerlinger, P. (2009). *How Birds Migrate* (Mechanicsburg, PA: Stackpole Books).

Kiepenheuer, J. (1978). Inversion of the magnetic field during transport: Its influence on the homing behavior of pigeons. In K. Schmidt-Koenig and W. T. Keeton (Eds.), pp. 135–142.

Kiepenheuer, J. (1982). The effect of magnetic anomalies on the homing behaviour of pigeons: An attempt to analyse the possible factors involved. In F. Papi and H. G. Wallraff (Eds.), pp. 120–128.

Kiepenheuer, J. (1986). A further analysis of the orientation behaviour of homing pigeons released within magnetic anomalies. In G. Maret, N. Boccara, and J. Kiepenheuer (Eds.), *Biophysical Effects of Steady Magnetic Fields* (Berlin, Germany: Springer-Verlag), pp. 148–153.

Kimchi, T., Etienne, A. S., and Terkel, J. (2004). A subterranean mammal uses the magnetic compass for path integration. *Proceedings of the National Academy of Sciences* 101, 1105–1109.

Kirschvink, J. L., Dizon, A. E., and Westphal, J. A. (1986). Evidence from strandings for geomagnetic sensitivity in cetaceans. *Journal of Experimental Biology* 120, 1–24.

Kirschvink, J. L., and Gould, J. L. (1981). Biogenic magnetite as a basis for magnetic field detection in animals. *BioSystems* 13, 181–201.

Klinowska, M. (1985). Cetacean live stranding sites related to geomagnetic topography. *Aquatic Mammals* 11, 27–32.

Kowalski, U., Wiltschko, R., and Füller, E. (1988). Normal fluctuations of the geomagnetic field may affect initial orientation in pigeons. *Journal of Comparative Physiology A: Neuroethology, Sensory, Neural, and Behavioral Physiology* 163, 593–600.

Kramer, G. (1959). Recent experiments on bird orientation. *Ibis* 101, 399–416.

Larkin, T. S., and Keeton, W. T. (1976). Bar magnets mask the effect of normal magnetic disturbances on pigeon orientation. *Journal of Comparative Physiology* 110, 227–231.

Lednor, A. J., and Walcott, C. (1988). Orientation of homing pigeons at magnetic anomalies. *Behavioral Ecology and Sociobiology* 22, 3–8.

Lohmann, K. J. (2007). Sea turtles: Navigating with magnetism. *Current Biology* 17, R102–R104.

Lohmann, K. J., Cain, S. D., Dodge, S. A., and Lohmann, C. M. F. (2001). Regional magnetic fields as navigational markers for sea turtles. *Science* 294, 364–366.

Lohmann, K. J., and Lohmann, C. M. F. (1994). Detection of magnetic inclination angle by sea turtles: A possible mechanism for determining latitude. *Journal of Experimental Biology* 194, 23–32.

Lohmann, K. J., and Lohmann, C. M. F. (1996). Detection of magnetic field intensity by sea turtles. *Nature* 380, 59–61.

Lohmann, K. J., Lohmann, C. M. F., Ehrhart, et al. (2004). Geomagnetic map used in sea turtle navigation. *Nature* 428, 909.

Lohmann, K. J., Lohmann, C. M. F., and Putman, N. F. (2007). Magnetic maps in animals: Nature's GPS. *Journal of Experimental Biology* 210, 3697–3705.

Lohmann, K. J., Pentcheff, N. D., Nevitt, G. A., et al. (1995). Magnetic orientation of spiny lobsters in the ocean: Experiments with undersea coil systems. *Journal of Experimental Biology* 198, 2041–2048.

Luschi, P., del Seppia, C., Crosio, E., and Papi, F. (1996). Pigeon homing: Evidence against reliance on magnetic information picked up en route to release sites. *Proceedings of the Royal Society of London B: Biological Sciences* 263, 1219–1224.

Malakoff, D. (1999). Following the scent of avian navigation. *Science* 286, 704–706.

McCaffery, B. J. (2008). On scimitar wings. *Birding* 40 (5), 50–59.

Mehlhorn, J., Haastert, B., and Rehkämper, G. (2010). Asymmetry of different brain structures in homing pigeons with and without navigational experience. *Journal of Experimental Biology* 213, 2219–2224.

Meyer, C. G., Holland, K. N., and Papastamatiou, Y. P. (2005). Sharks can detect changes in the geomagnetic field. *Journal of the Royal Society Interface* 2, 129–130.

Michener, M., and Walcott, C. (1967). Homing of single pigeons—Analysis of tracks. *Journal of Experimental Biology* 47, 99–131.

Moore, B. (1980). Is the homing pigeon's map magnetic? *Nature* 285, 69–70.

Mora, C. V., Davison, M., Wild, J. M., and Walker, M. M. (2004). Magnetoreception and its trigeminal mediation in the homing pigeons. *Nature* 432, 508–511.

Morreale, S. J., Standora, E. A., Spotila, J. R., and Paladino, F. V. (1996). A migration corridor for sea turtles. *Nature* 384, 319–320.

Mouritsen, H., Feenders, G., Liedvogel, M., and Kropp, W. (2004). Migratory birds use head scans to detect the direction of the earth's magnetic field. *Current Biology* 14, 1946–1949.

Munro, U., Munro, J., and Phillips, J. B. (1997). Evidence for a magnetite-based navigational "map" in birds. *Naturwissenschaften* 84, 26–28.

Munro, U., Munro, J., Phillips, J. B., and Wiltschko, W. (1997). Effect of wavelength of light and pulse magnetization on different magnetoreception systems in a migrating bird. *Australian Journal of Zoology* 45, 189–198.

Ottosson, U., Sandberg, R., and Pettersson, J. (1990). Orientation cage and release experiments with migratory wheatears in Scandinavia and Greenland: The importance of visual cues. *Ethology* 86, 57–70.

Papi, F. (1982). Olfaction and homing in pigeons. In F. Papi and H. G. Wallraff (Eds.), pp. 149–159.

Papi, F. (1986). Pigeon navigation: Solved problems and open questions. *Monitore Zoologico Italiano* 20, 471–517.

Papi, F. (1991). Olfactory navigation. In P. Berthold (Ed.), *Orientation in Birds* (Basel, Switzerland: Birkhäuser Verlag), pp. 52–85.

Papi, F. (1992). *Animal Homing* (London, United Kingdom: Chapman and Hall).

Papi, F., Fiore, L., Fiaschi, V., and Benvenuti, S. (1972). Olfaction and homing in pigeons. *Monitore Zoologico Italiano* 6, 85–95.

Papi, F., Ioalè, P., Fiaschi, V., et al. (1978). Pigeon homing: Cues detected during the outward journey influence initial orientation. In K. Schmidt-Koenig and W. T. Keeton (Eds.), pp. 65–77.

Papi, F., Keeton, W. T., Brown, A. I., and Benvenuti, S. (1978). Do American and Italian pigeons rely on different homing mechanisms? *Journal of Comparative Physiology* 128, 303–317.

Papi, F., Luschi, P. (1996). Pinpointing "Isla Meta": The case of sea turtles and albatrosses. *Journal of Experimental Biology* 199, 65–71.

Papi, F., and Wallraff, H. G. (Eds.). (1982). *Avian Navigation* (Berlin, Germany: Springer-Verlag).

Phillips, J. B. (1986). Two magnetoreception pathways in a migratory salamander. *Science* 233, 765–767.

Phillips, J. B. (1987). Laboratory studies of homing orientation in the eastern red-spotted newt. *Journal of Experimental Biology* 131, 215–229.

Phillips, J. B. (1996). Magnetic navigation. *Journal of Theoretical Biology* 180, 309–319.

Phillips, J. B., Adler, K., and Borland, S. C. (1995). True navigation by an amphibian. *Animal Behaviour* 50, 855–858.

Phillips, J. B., Freake, M. J., Fischer, J. H., and Borland, S. C. (2002). Behavioral titration of a magnetic map coordinate. *Journal of Comparative Physiology A: Neuroethology, Sensory, Neural, and Behavioral Physiology* 188, 157–160.

Phillips, J. B., and Waldvogel, J. A. (1982). Reflected light cues generate the short-term deflector-loft effect. In F. Papi and H. G. Wallraff (Eds.), pp. 190–202.

Putnam, N. F., Endres, C. S., Lohmann, C. M. F., and Lohmann, K. J. (2011). Longitude perception and bicoordinate magnetic maps in sea turtles. *Current Biology* 21, 463–466.

Putnam, N. F., and Lohmann, K. J. (2008). Compatibility of magnetic imprinting and secular variation. *Current Biology* 18, R596–R597.

Sandberg, R. (1991). Sunset orientation of robins with different fields of sky vision. *Behavioral Ecology and Sociobiology* 28, 77–83.

Schmidt-Koenig, K. (1975). *Migration and Homing in Animals* (Berlin, Germany: Springer-Verlag).

Schmidt-Koenig, K. (1979). *Avian Orientation and Navigation* (New York, NY: Academic Press).

Schmidt-Koenig, K. (1985). Hypothesen und Argumente zum Navigationsvermögen der Vögel. *Journal of Ornithology* 126, 237–252.

Schmidt-Koenig, K. (1987). Bird navigation: Has olfactory orientation solved the problem? *Quarterly Review of Biology* 62, 31–47.

Schmidt-Koenig, K., and Keeton, W. T. (Eds.). (1978). *Animal Migration, Navigation, and Homing* (Berlin, Germany: Springer-Verlag).

Schmidt-Koenig, K., and Walcott, C. (1978). Tracks of pigeons homing with frosted lenses. *Animal Behaviour* 26, 480–486.

Schreiber, B., and Rossi, O. (1976). Correlation between race arrivals of homing pigeons and solar activities. *Bollettino di Zoologia* 43, 317–320.

Schreiber, B., and Rossi, O. (1978). Correlation between magnetic storms due to solar spots and pigeon homing performances. *IEEE Transactions on Magnetics* 14, 961–963.

Seachrist, L. (1994). Sea turtles master migration with magnetic memories. *Science* 264, 661–662.

Semm, P., and Beason, R. C. (1990). Responses to small magnetic variations by the trigeminal system of the bobolink. *Brain Research Bulletin* 25, 735–740.

Stutchbury, B. J. M., Tarof, S. A., Done, T., et al. (2009). Tracking long-distance songbird migration by using geolocators. *Science* 323, 896.

Thorup, K., Bisson, I. A., Bowlin, M. S., et al. (2007). Evidence for a navigational map stretching across the continental U.S. in a migratory songbird. *Proceedings of the National Academy of Sciences* 104, 18115–18119.

Thorup, K., and Holland, R. A. (2009). The bird GPS—Long-range navigation in migrants. *Journal of Experimental Biology* 212, 3597–3604.

Viguier, C. (1882). Le sens de l'orientation et ses organs chez les animaux et chez l'homme. *Revue Philosophique de la France et de l'Etranger* 14, 1–36.

Visalberghi, E., and Alleva, E. (1979). Magnetic influence on pigeon homing. *Biological Bulletin* 156, 246–256.

Wagner, G. (1970). Verfolgung von Brieftauben im Helikopter. *Revue Suisse de Zoologie* 77, 39–60.

Wagner, G. (1974). Verfolgung von Brieftauben im Helikopter II. *Revue Suisse de Zoologie* 80, 727–750.

Wagner, G. (1976). Das Orientierungsverhalten von Brieftauben im erdmagnetisch gestörten Gebiete des Chasseral. *Revue Suisse de Zoologie* 83, 883–890.

Wagner, G. (1983). Natural geomagnetic anomalies and homing in pigeons. *Comparative Biochemistry and Physiology A: Comparative Physiology* 76, 691–700.

Walcott, C. (1974). The homing of pigeons. *American Scientist* 62, 542–552.

Walcott, C. (1978). Anomalies in the earth's magnetic field increase the scatter of pigeons' vanishing bearings. In K. Schmidt-Koenig and W. T. Keeton (Eds.), pp. 143–151.

Walcott, C. (1980). Magnetic orientation in homing pigeons. *IEEE Transactions on Magnetics* 16, 1008–1013.

Walcott, C. (1992). Pigeons at magnetic anomalies: The effects of loft location. *Journal of Experimental Biology* 170, 127–141.

Walcott, C., Gould, J. L., and Kirschvink, J. L. (1979). Pigeons have magnets. *Science* 205, 1027–1029.

Walcott, C., Gould, J. L., and Lednor, A. J. (1988). Homing of magnetized and demagnetized pigeons. *Journal of Experimental Biology* 134, 27–41.

Walcott, C., and Schmidt-Koenig, K. (1973). The effect on pigeon homing of anesthesia during displacement. *Auk* 90, 281–286.

Waldvogel, J. A., and Phillips, J. B. (1982). Pigeon homing: New experiments involving permanent-resident deflector-loft birds. In F. Papi and H. G. Wallraff (Eds.), pp. 179–189.

Walker, M. M. (1998). On a wing and a vector: A model for magnetic navigation by homing pigeons. *Journal of Theoretical Biology* 192, 341–349.

Walker, M. M. (1999). Magnetic position determination by homing pigeons. *Journal of Theoretical Biology* 197, 271–276.

Walker, M. M. (2008). A model for encoding of magnetic field intensity by magnetite-based magnetoreceptor cells. *Journal of Theoretical Biology* 250, 85–91.

Walker, M. M., Dennis, T. E., and Kirschvink, J. L. (2002). The magnetic sense and its use in long distance navigation by animals. *Current Opinion in Neurobiology* 12, 735–744.

Wallraff, H. G. (1960). Über Zusammenhänge des Heimkehrverhaltens von Brieftauben mit meteorologischen und geophysikalischen Faktoren. *Zeitschrift für Tierpsychologie* 17, 82–113.

Wallraff, H. G. (1966). Über die Anfangsorientierung von Brieftauben uter geschlossener Wolkendecke. *Journal of Ornithology* 107, 326–336.

Wallraff, H. G. (1966). Über die Heimfindeleistungen von Brieftauben nach Haltung in verschiedenartig abgeschirmten Volieren. *Zeitschrift für Vergleichende Physiologie* 52, 215–259.

Wallraff, H. G. (1980). Does pigeon homing depend on stimuli perceived during displacement? I: Experiments in Germany. *Journal of Comparative Physiology* 139, 193–201.

Wallraff, H. G. (1981). The olfactory component of pigeon navigation. *Journal of Comparative Physiology* 143, 411–422.

Wallraff, H. G. (1982). Homing to Würzburg: An interim report on long-term analyses of pigeon navigation. In F. Papi and H. G. Wallraff (Eds.), pp. 211–221.

Wallraff, H. G. (1999). The magnetic map of homing pigeons: An evergreen phantom. *Journal of Theoretical Biology* 197, 265–269.

Wallraff, H. G. (2004). Avian olfactory navigation: Its empirical foundation and conceptual state. *Animal Behaviour* 67, 189–204.

Wallraff, H. G. (2005). *Avian Navigation: Pigeon Homing as a Paradigm* (Berlin, Germany: Springer-Verlag).

Westby, G. W. M., and Partridge, K. J. (1986). Human homing: Still no evidence despite geomagnetic controls. *Journal of Experimental Biology* 120, 325–331.

Wiltschko, R. (1991). Role of experience in avian navigation and homing. In P. Berthold (Ed.), *Orientation in Birds* (Basel, Switzerland: Birkhäuser Verlag), pp. 250–269.

Wiltschko, R. (1996). The function of olfactory input in pigeon orientation: Does it provide navigational information or play another role? *Journal of Experimental Biology* 199, 113–119.

Wiltschko, R., Schiffner, I., and Wiltschko, W. (2009). A strong magnetic anomaly affects pigeon navigation (1996). *Journal of Experimental Biology* 212, 2983–2990.

Wiltschko, R., and Wiltschko, W. (1985). Pigeon homing: Change in navigational strategy during ontogeny. *Animal Behaviour* 33, 583–590.

Wiltschko, R., and Wiltschko, W. (1995). *Magnetic Orientation in Animals* (Berlin, Germany: Springer-Verlag).

Wiltschko, R., Wiltschko, W., and Keeton, W. T. (1978). Effect of outward journey in an altered magnetic field on the orientation of young homing pigeons. In K. Schmidt-Koenig and W. T. Keeton (Eds.), pp. 152–161.

Wiltschko, W., Munro, U., Beason, R. C., et al. (1994). A magnetic pulse leads to a temporary deflection in the orientation of migratory birds. *Experientia* 50, 697–700.

Wiltschko, W., Munro, U., and Wiltschko, R. (1997). Magnetoreception in migratory birds: Light-mediated and magnetite-mediated processes? In *Orientation & Navigation—Birds, Humans, and Other Animals* (Oxford, United Kingdom: Royal Institute of Navigation), pp. 1.1–1.9.

Wiltschko, W., and Wiltschko, R. (1981). Disorientation of inexperienced homing pigeons after transportation in total darkness. *Nature* 291, 433–434.

Wiltschko, W., and Wiltschko, R. (1995). Migratory orientation of European robins is affected by the wavelength of light as well as by a magnetic pulse. *Journal of Comparative Physiology A: Neuroethology, Sensory, Neural, and Behavioral Physiology* 177, 363–369.

Wiltschko, W., and Wiltschko, R. (1996). Magnetic orientation in birds. *Journal of Experimental Biology* 199, 29–38.

Windsor, D. M. (1975). Regional expression of directional preferences by experienced homing pigeons. *Animal Behaviour* 23, 335–343.

Wu, L.-Q., and Dickman, D. (2011). Magnetoreception in an avian brain in part mediated by inner ear lagena. *Current Biology* 21, 418–423.

Zimmer, C. (2010, May 25). 7000 miles nonstop, and no pretzels. *New York Times*, pp. D1, D3.

Chapter 8

Anonymous. (1974, June 24). Another ice age? *Time* 103 (25), 86.

Bearhop, S., Fiedler, W., Furness, W., et al. (2005). Assortative mating as a mechanism for rapid evolution of a migratory divide. *Science* 310, 502–504.

Berthold, P., Helbig, A. J., Mohr, G., and Querner, U. (1992). Rapid microevolution of migratory behavior in a wild bird species. *Nature* 360, 668–670.

Both, C. (2010). Flexibility of timing of avian migration to climate change masked by environmental constraints en route. *Current Biology* 20, 243–248.

Bridge, E. S., Kelly, J. F., Bjornen, P. E., et al. (2010). Effects of nutritional condition on spring migration: Do migrants use resource availability to keep pace with a changing world? *Journal of Experimental Biology* 213, 2424–2429.

Bruderer, B., and Salewski, V. (2008). Evolution of bird migration in a biogeographical context. *Journal of Biogeography* 35, 1951–1959.

Butler, C. (2003). Disproportionate effect of global warming on the arrival dates of short-distance migratory birds in North America. *Ibis* 145, 484–495.

Coppack, T., Pulido, F., Czisch, M., et al. (2003). Photoperiodic response may facilitate adaptation to climatic change in long-distance migratory birds. *Proceedings of the Royal Society of London B: Biological Sciences* 270, S43–S46.

Coppack, T., Tindemans, I., Czisch, M., et al. (2008). Can long-distance migratory birds adjust to the advancement of spring by shortening migration distance? The response of the pied flycatcher to latitudinal photoperiodic variation. *Global Change Biology* 14, 2516–2522.

Cox, G. W. (2010). *Bird Migration and Global Change* (Washington, DC: Island Press).

Crick, H. Q. P., and Sparks, T. H. (1999). Climate change related to egg-laying trends. *Nature* 399, 423–423.

Crowley, T. J., and Hyde, W. T. (2008). Transient nature of late Pleistocene climate variability. *Nature* 456, 226–230.

Davis, L. A., Roalson, E. H., Cornell, K. L., et al. (2006). Genetic divergence and migration patterns in a North American passerine bird: Implications for evolution and conservation. *Molecular Ecology* 15, 2141–2152.

Filippi-Codaccioni, O., Moussus, J. P., Urcun, J. P., and Jiguet, F. (2010). Advanced departure dates in long-distance migratory raptors. *Journal of Ornithology* 151, 687–694.

Fontaine J. J., Decker, K. L., Skagen, S. K., and van Riper, C. (2009). Spatial and temporal variation in climate change: A bird's eye view. *Climate Change* 97, 305–311.

Gienapp, P., Leimu, R., and Merilä, L. (2007). Responses to climate change in avian migration time: Microevolution vs. phenotypic plasticity. *Climate Research* 35, 25–35.

Gore, A. (2006). *An Inconvenient Truth* (New York, NY: Rodale Press).

Gould, J. L., and Gould, C. G. (1996). *Sexual Selection* (New York, NY: W. H. Freeman).

Graedel, T. E., and Crutzen, P. J. (1995). *Atmosphere, Climate, and Change* (New York, NY: W. H. Freeman).

Gwynne, P. (1975, April 28). The cooling world. *Newsweek*, p. 64.

Halupka, L., Dyrcz, A., and Borowiec, M. (2008). Climate change affects breeding of reed warblers. *Journal of Avian Biology* 39, 95–100.

Helbig, A. J. (1991). Inheritance of migratory direction in a bird species: A cross-breeding experiment with SE- and SW-migrating blackcaps. *Behavioral Ecology and Sociobiology* 28, 9–12.

Helbig, A. J. (1996). Genetic basis, mode of inheritance, and evolutionary changes of migratory directions in Palearctic warblers. *Journal of Experimental Biology* 199, 49–55.

Helm, B., Schwabl, I., and Gwinner, E. (2009). Circannual basis of geographically distinct bird schedules. *Journal of Experimental Biology* 212, 1259–1269.

Hubalek, Z., Capek, M. (2008). Migration distance and the effect of North Atlantic Oscillation on the spring arrival of birds in central Europe. *Folia Zoologica* 57, 212–220.

Jonzén, N., Lindén, A., Ergon, T., et al. (2006). Rapid advance of spring arrival dates in long-distance migratory birds. *Science* 312, 1959–1961.

Kerlinger, P. (2009). *How Birds Migrate* (Mechanicsburg, PA: Stackpole Books).

La Sorte, F. A., and Jetz, W. (2010). Avian distributions under climate change: Towards improved projections. *Journal of Experimental Biology* 213, 862–869.

Lomborg, L. (2007). *Cool It: The Skeptical Environmentalist's Guide to Global Warming* (New York, NY: Knopf).

Louchart, A. (2008). Emergence of long distance bird migrations: A new model integrating global climate changes. *Naturwissenschaften* 95, 1109–1119.

Lyon, B. E., Chaine, A. S., and Winkler, D. W. (2008). A matter of timing. *Science* 321, 1051–1052.

Markham, A., Dudley, N., and Stolton, S. (1993). *Some Like It Hot: Climate Change, Biodiversity, and the Survival of Species* (Gland, Switzerland: WWF International).

Møller, A. P., Fiedler, W., and Berthold, P. (2004). *Birds and Climate Change* (New York, NY: Academic Press).

Møller, A. P., Flensted-Jensen, E., Klarborg, K., et al. (2010). Climate change affects the duration of the reproductive season in birds. *Journal of Animal Ecology* 79, 777–784.

Møller, A. P., Rubolini, D., and Lehikoinen, E. (2008). Populations of migratory bird species that did not show a phenological response to climate change are declining. *Proceedings of the National Academy of Sciences* 105, 16195–16200.

Mueller, J. C., Pulido, F., and Kempenaers, B. (2011). Identification of a gene associated with avian migratory behaviour. *Proceedings of the Royal Society B: Biological Sciences.* http://rspb.royalsocietypublishing.org/content/early/2011/02/11/rspb.2010.2567.full?sid=3d519182-ac1f-4222-9ed5-90848f5f84b0 (accessed 25 July 2011).

Newson, S. E., Dulvy, N., Hays, G. C., and Houghton, J. D. R. (2008). *Indicators of the Impact of Climate Change on Migratory Species* (London, United Kingdom: British Trust for Ornithology).

Nyberg, J., Malmgren, B. A., Winter, A., et al. (2007). Low Atlantic hurricane activity in the 1970s and 1980s compared to the past 270 years. *Nature* 447, 698–701.

Pearson, R. G. (2006). Climate change and the migration capacity of species. *Trends in Ecology & Evolution* 21, 111–113.

Petit, J. R., Jouzel, J., Raynaud, D., et al. (1999). Climate and atmospheric history of the past 420,000 years from the Vostok ice core, Antarctica. *Nature* 399, 429–436.

Philander, S. G. (1998). *Is the Temperature Rising?* (Princeton, NJ: Princeton University Press).

Pulido, F. (2007). The genetics and evolution of avian migration. *BioScience* 57, 165–174.

Pulido, F., Berthold, P., Mohr, G., and Querner, U. (2001). Heritability of the timing of autumn migration in a natural bird population. *Proceedings of the Royal Society of London B: Biological Sciences* 268, 953–959.

Pulido, F., and Widmer, M. (2005). Are long-distance migrants constrained in their evolutionary response to environmental change? Causes of variation in the timing of autumn migration in a blackcap and two garden warbler populations. In U. Bauchinger, W. Goymann, and S. Jenni-Eiermann (Eds.), *Bird Hormones and Bird Migrations* (New York, NY: New York Academy of Sciences), pp. 228–241.

Rainio, K., Laaksonen, T., Ahola, M., et al. (2006). Climatic responses in spring migration of boreal and arctic birds in relation to wintering area and taxonomy. *Journal of Avian Biology* 37, 507–515.

Reilly, J. R., and Reilly, R. J. (2009). Bet-hedging and the orientation of juvenile passerines in fall migration. *Journal of Animal Ecology* 78, 990–1001.

Rivalan, P., Frederiksen, M., Lois, G., and Julliard, R. (2007). Contrasting responses of migration strategies in two European thrushes to climate change. *Global Change Biology* 13, 275–287.

Rolshausen, G., Segelbacher, G., Hobson, K. A., and Schaefer H. M. (2009). Contemporary evolution of reproductive isolation and phenotypic divergence in sympatry along a migratory divide. *Current Biology* 19, 2097–2101.

Rubolini, D., Saino, N., and Møller, A. P. (2010). Migratory behavior constrains the phenological response of birds to climate change. *Climate Research* 42, 45–55.

Smallegange, I. M., Fiedler, W., Koppen, U., et al. (2010). Tits on the move:

Exploring the impact of environmental change on blue tit and great tit migration distance. *Journal of Animal Ecology* 79, 350–357.

Sparks, T. H., Bairlein, F., Bojarinova, J. G., et al. (2005). Examining the total arrival distribution of migratory birds. *Global Change Biology* 11, 22–30.

Sullivan, W. (1975, May 21). Scientists ask why world climate is changing; a major cooling widely considered to be inevitable. *New York Times*, p. 45.

Vagg, R., and Hepworth, H. (Eds.). (2007). *Migratory Species and Climate Change: Impacts of a Changing Environment on Wild Animals* (Bonn, Germany: United Nations Environment Programme).

Visser, M. E., Perdeck, A. C., and van Balen, J. H. (2009). Climate change leads to decreasing bird migration distances. *Global Change Biology* 15, 1859–1865.

Wilcove, D. S. (2008). *No Way Home* (Washington, DC: Island Press), p. 210.

Illustration Credits

Page 4: Original drawing.
Page 7: National Aeronautics and Space Administration.
Page 9: Authors' photograph.
Page 12: Redrawn from A. B. Bochdansky and S. M. Bollens (2004), Relevant scales in zooplankton ecology: Distribution, feeding, and reproduction of the copepod *Acartia hudsonica* in response to thin layers of the diatom *Skeletonema costatum*, *Limnology and Oceanography* 49, 625–636. Copyright 2004 by the American Society of Limnology and Oceanography, Inc.
Page 14: Original drawing.
Page 15: Original drawing.
Page 20: Original drawing.
Page 21: Original drawing.
Pages 24–25: Original drawing.
Page 27: Redrawn from a US Geological Survey coastal map.
Page 27: Original drawing.
Page 30: Original drawing.
Page 30: Redrawn from Free Software Foundation, Inc.
Page 41: Redrawn from P. J. DeCoursey (1961), Effect of light on the circadian activity rhythm of the flying squirrel, *Glaucomys volans*, *Zeitschrift für Vergleichende Physiologie* 44, 331–354.
Page 45: Illustration taken from *The Honey Bee* by James L. Gould and Carol Grant Gould. Copyright © 1988 by Scientific American Library.

Reprinted by permission of Henry Holt and Company, LLC (New York, NY).

Page 49: Redrawn from M. L. Walls and J. E. Layne (2009), Direct evidence for distance measurement via flexible stride integration in the fiddler crab, *Current Biology* 19, 25–29, © 2009, with permission from Elsevier.

Page 53: Original drawing.

Page 58: Original drawing.

Page 60: Original drawing.

Page 62: Original drawing.

Page 63: Redrawn from *Biological Science*, Sixth Edition by James L. Gould and William T. Keeton. Copyright © 1996, 1993, 1986, 1980, 1979, 1978, 1967 by W. W. Norton & Company, Inc. Used by permission of W. W. Norton &Company, Inc. (New York, NY).

Page 66: Original drawing.

Page 72: Original drawing.

Page 76: Authors' photographs.

Page 79: Redrawn from *Ethology: The Mechanisms and Evolution of Behavior* by James L. Gould. Copyright © 1982 by James L. Gould. Used by permission of W. W. Norton & Company, Inc. (New York, NY).

Page 81: Original drawing.

Page 82: Redrawn from M. L. Brines and J. L. Gould (1979), Bees have rules, *Science* 206, 571–573, with permission from The American Association for the Advancement of Science.

Page 89: Redrawn from J. L. Gould (1980), Sun compensation by bees, *Science* 207, 545–547, with permission from The American Association for the Advancement of Science.

Page 90: Redrawn from J. L. Gould (1984), Processing of sun-azimuth information by honey bees, *Animal Behaviour* 32, 149–152, © 1984, with permission from Elsevier.

Page 92: Authors' photographs.

Page 93: Redrawn from *Ethology: The Mechanisms and Evolution of Behavior* by James L. Gould. Copyright © 1982 by James L. Gould. Used by permission of W. W. Norton & Company, Inc. (New York, NY).

Page 94: Redrawn from *Ethology: The Mechanisms and Evolution of Behavior* by James L. Gould. Copyright © 1982 by James L. Gould. Used by permission of W. W. Norton & Company, Inc. (New York, NY).

Page 95: Redrawn from *Ethology: The Mechanisms and Evolution of Behav-*

ior by James L. Gould. Copyright © 1982 by James L. Gould. Used by permission of W. W. Norton & Company, Inc. (New York, NY).

Page 98: Redrawn from F. C. Dyer and J. L. Gould (1980), Honey bee orientation: A backup system for cloudy days, *Science* 214, 1041–1042, with permission from The American Association for the Advancement of Science.

Page 99: Redrawn from F. C. Dyer and J. L. Gould (1980), Honey bee orientation: a backup system for cloudy days, *Science* 214, 1041–1042, with permission from The American Association for the Advancement of Science.

Page 101: Redrawn from http://en.wikipedia.org/wiki/File:Geomagnetisme.svg, user JrPol.

Page 102: Redrawn from M. Lindauer and H. Martin (1972), Magnetic effects on dancing bees, in S. R. Galler, et al. (Eds.), *Animal Orientation & Navigation* (Washington, DC: US Government Printing Office), pp. 559–567.

Page 104: Illustration taken from *The Honey Bee* by James L. Gould and Carol Grant Gould. Copyright © 1988 by Scientific American Library. Reprinted by permission of Henry Holt and Company, LLC (New York, NY).

Page 106: Redrawn from a map by US Geological Survey.

Page 108: Redrawn from J. L. Gould (2010), Magnetoreception, *Current Biology* 20, R431–435, © 2010, with permission from Elsevier.

Page 111: Original drawing.

Page 113: Redrawn from J. L. Gould (2010), Magnetoreception, *Current Biology* 20, R431–R435, © 2010, with permission from Elsevier.

Page 119: Redrawn from W. Wiltschko (1968), Über den Einfluß statischer Magnetfelder auf die Zugorientierung der Rotkehlchen. *Zeitschrift für Tierpsychologie* 25, 537–558.

Page 119: Original drawing.

Page 122: Redrawn from B. Elsner (1978), Accurate measurements of the initial tracks of homing pigeons, in K. Schmidt-Koenig and W. T. Keeton (Eds.), *Animal Migration, Navigation, and Homing* (Berlin, Germany: Springer-Verlag), pp. 194–198.

Page 123: Original drawing.

Page 124: Original drawing.

Page 124: Redrawn from W. T. Keeton (1974), The mystery of pigeon homing, *Scientific American* 231 (6), 96–104, with the permission of the estate of Bunji Tagawa.

Page 125: Adapted with permission from M. Michener and C. Walcott (1967), Homing of single pigeons—Analysis of tracks, *Journal of Experimental Biology* 47, 99–131.

Page 127: Redrawn from K. Schmidt-Koenig and C. Walcott (1978), Tracks of pigeons homing with frosted lenses, *Animal Behaviour* 26, 480–486, © 1978, with permission from Elsevier.

Page 135: Original drawing.

Page 136: Redrawn from W. T. Keeton (1969), Orientation by pigeons: Is the sun necessary? *Science* 165, 922–928, with permission from The American Association for the Advancement of Science.

Page 139: Original drawing.

Page 140: © David Malin, used with permission.

Page 141: Original drawing.

Page 143: Redrawn from R. Muheim and S. Åkesson (2002), Clock-shift experiments with savannah sparrows at high northern latitudes, *Behavioral Ecology and Sociobiology* 51, 394–401.

Page 144: Redrawn from W. T. Keeton (1971), Magnets interfere with pigeon homing, *Proceedings of the National Academy of Sciences* 68, 102–106, used with the permission of Barbara "Bobby" O. Keeton (Mrs. W. T. Keeton).

Page 145: Redrawn from C. Walcott and R. P. Green (1974), Orientation of homing pigeons altered by a change in the direction of an applied magnetic field, *Science* 184, 180–182, with permission from The American Association for the Advancement of Science.

Page 151: Original drawing.

Page 156: Based on a description in W. Köhler (1927), *The Mentality of Apes* (New York, NY: Harcourt Brace).

Page 160: Illustration taken from *The Animal Mind* by James L. Gould and Carol Grant Gould. Copyright © 1994 by Scientific American Library. Reprinted by permission of Henry Holt and Company, LLC (New York, NY).

Page 161: Original drawing.

Page 163: Redrawn from N. Chapuis and P. Scardigli (1993), Shortcut ability in hamsters, *Animal Learning and Behavior* 21, 255–265, with permission of the Psychonomic Society.

Page 164: Original drawing.

Page 165: Redrawn from J. L. Gould (1986), The locale map of honey bees: Do insects have cognitive maps? *Science* 232, 861–863, with permission from The American Association for the Advancement of Science.

Page 166: Redrawn from M. S. Tarsitano and R. R. Jackson (1994), Jumping spiders make predatory detours requiring movement away from prey, *Behaviour* 131, 65–73, with permission of Koninklijke Brill NV.

Page 167: Illustration taken from *The Honey Bee* by James L. Gould and Carol Grant Gould. Copyright © 1988 by Scientific American Library. Reprinted by permission of Henry Holt and Company, LLC (New York, NY).

Page 171: Redrawn from H. A. Isack and H.-U. Reyer (1989), Honeyguides and honey gathers, *Science* 243, 1343–1346, with permission from The American Association for the Advancement of Science.

Page 172: Illustration taken from *The Honey Bee* by James L. Gould and Carol Grant Gould. Copyright © 1988 by Scientific American Library. Reprinted by permission of Henry Holt and Company, LLC (New York, NY).

Page 174: Redrawn from N. Tinbergen and W. Kruyt (1938), Über die Orientierung des Bienenwolfes III: Die Bevorzugung bestimmter Wegmarken, *Zeitschrift für Vergleichende Physiologie* 25, 292–234.

Page 175: Redrawn from J. L. Gould (1987), Landmark learning by honey bees, *Animal Behaviour* 35, 26–34, © 1987, with permission from Elsevier.

Page 177: Redrawn from J. L. Gould (1987), Landmark learning in honey bees, *Animal Behaviour* 35, 26–34, © 1987, with permission from Elsevier.

Page 179: Redrawn from J. L. Souman, I. Frissen, M. N. Sreenlvasa, and M. O. Ernst (2009), Walking straight into circles, *Current Biology* 19, 1538–1542, © 2009, with permission from Elsevier.

Page 180: Redrawn from J. L. Souman, I. Frissen, M. N. Sreenlvasa, and M. O. Ernst (2009), Walking straight into circles, *Current Biology* 19, 1538–1542, © 2009, with permission from Elsevier.

Page 182: Redrawn from J. L. Souman, I. Frissen, M. N. Sreenlvasa, and M. O. Ernst (2009), Walking straight into circles, *Current Biology* 19, 1538–1542, © 2009, with permission from Elsevier.

Page 186: Redrawn from Gill, R. E., Tibbitts, T. L., Douglas, D. C., et al. (2009), Extreme endurance flights by landbirds crossing the Pacific Ocean: Ecological corridor rather than barrier?, *Proceedings of the Royal Society B: Biological Sciences* 276, 447–457, with permission from Highwire Press.

Page 191: Redrawn from N. Chernetsov, D. Kishkinev, and H. Mouritsen (2008), A long-distance avian migrant compensates for longitudinal displacement during spring migration, *Current Biology* 18, 188–190, © 2008, with permission from Elsevier.

Page 192: Redrawn from K. Thorup, I. A. Bisson, M. S. Bowlin, et al. (2007), Evidence for a navigational map stretching across the continental U.S. in a migratory songbird, *Proceedings of the National Academy of Sciences* 104, 18115–18119, © 2007, with permission from the National Academy of Sciences, USA.

Page 193: Redrawn from K. Thorup, I. A. Bisson, M. S. Bowlin, et al. (2007), Evidence for a navigational map stretching across the continental U.S. in a migratory songbird, *Proceedings of the National Academy of Sciences* 104, 18115–18119, © 2007, with permission from the National Academy of Sciences, USA.

Page 195: Original drawing.

Page 196: Original drawing.

Page 198: Original drawing.

Page 200: Original drawing based on photographs in H. G. Wallraff (1966), Über die Heimfindeleistungen von Brieftauben nach Haltung in verschiedenartig abgeschirmten Volieren, *Zeitschrift für Vergleichende Physiologie* 52, 215–259.

Page 203: Redrawn from F. Papi (1982), Olfaction and homing in pigeons, in F. Papi and H. G. Wallraff (Eds.), *Avian Navigation* (Berlin, Germany: Springer-Verlag), pp. 149–159.

Page 205: Redrawn from P. Ioalè, M. Nozzolini, and F. Papi (1990), Pigeons do extract directional information from olfactory stimuli, *Behavioral Ecology and Sociobiology* 26, 301–305.

Page 207: Redrawn from J. L. Gould (1982), The map sense of pigeons, *Nature* 296, 205–211, © 1982, by permission of Nature Publishing Group.

Page 208: Redrawn from W. T. Keeton (1973), Release site bias as a possible guide to the "map" component in pigeon homing, *Journal of Comparative Physiology* 86, 1–16.

Page 209: Redrawn from D. M. Windsor (1975), Regional expression of directional preferences by experienced homing pigeons, *Animal Behaviour* 23, 335–343, © 1975, with permission from Elsevier.

Page 211: Redrawn from *Ethology: The Mechanisms and Evolution of Behavior* by James L. Gould. Copyright © 1982 by James L. Gould. Used by permission of W. W. Norton & Company, Inc. (New York, NY).

Page 211: Redrawn from C. Walcott (1978), Anomalies in the earth's magnetic field increase the scatter of pigeons' vanishing bearings, in

K. Schmidt-Koenig and W. T. Keeton (Eds.), *Animal Migration, Navigation, and Homing* (Berlin, Germany: Springer-Verlag), pp. 143–151.

Page 212: Original drawing based on US Geological Survey maps.

Page 215: Redrawn from J. B. Phillips, M. J. Freake, J. H. Fischer, and S. C. Borland (2002), Behavioral titration of a magnetic map coordinate. *Journal of Comparative Physiology A: Neuroethology, Sensory, Neural, and Behavioral Physiology* 188, 157–160.

Page 216: Redrawn from L. C. Boles and K. J. Lohmann (2003), True navigation and magnetic maps in spiny lobsters, *Nature* 421, 60–63, © 2003, by permission of Nature Publishing Group.

Page 219: Redrawn with permission from K. J. Lohmann and C. M. F. Lohmann (1994), Detection of magnetic inclination angle by sea turtles: A possible mechanism for determining latitude, *Journal of Experimental Biology* 194, 23–32.

Pages 220–221: Redrawn from K. J. Lohmann, S. D. Cain, S. A. Dodge, and C. M. F. Lohmann (2001), Regional magnetic fields as navigational markers for sea turtles, *Science* 294, 364–366, with permission from The American Association for the Advancement of Science.

Page 222: Redrawn from K. J. Lohmann, C. M. F. Lohmann, L. M. Ehrhart, D. A. Bagley, and K. Swing (2004), Geomagnetic map used in sea turtle navigation, *Nature* 428, 909, © 2004, by permission of Nature Publishing Group.

Page 224: Redrawn from R. R. Baker (1980), Goal orientation by blindfolded humans after long-distance displacement: Possible involvement of a magnetic sense, *Science* 210, 555–557, with permission from The American Association for the Advancement of Science.

Page 225: Redrawn from J. L. Gould and K. P. Able (1981), Human homing: An elusive phenomenon, *Science* 212, 1061–1063, with permission from The American Association for the Advancement of Science.

Page 229: Redrawn from images created by Robert A. Rohde, Global Warming Art Project.

Page 232: Original drawing.

Page 233: Redrawn from L. Halupka, A. Dyrcz, and M. Borowiec (2008), Climate change affects breeding of reed warblers, *Journal of Avian Biology* 39, 95–100, © 2008, with permission of John Wiley & Sons.

Page 235: Original drawing.

Page 236: Redrawn from P. Berthold, A. J. Helbig, G. Mohr, and U. Querner

(1992), Rapid microevolution of migratory behavior in a wild bird species, *Nature* 360, 668–670, © 1992, by permission of Nature Publishing Group.

Page 238: Redrawn from image created by Robert A. Rohde, Global Warming Art Project.

Index

Page numbers in *italics* refer to figures

anemotaxis, 13–15, *15*
anomaly, magnetic, 208–211, *209*, *211*
ants, desert, 50, 176; polarized-light use by, 91–92; sun compensation in, 73

bacteria, 13, *14*, 113–114
badgers, honey (ratels), 170–171, *171*
baited-air tests, 204, *205*
Baker, R., 223
bank swallows, 198
bar-tailed godwit, 185–186, *186*
bats, 147
beacons, use of, 128–133, 148, 180–181
bees, nocturnal, 75. *See also* honey bees
beetles, dung, 95–96
Behaviorism, 11, 157–159
Bermuda, 1–10, *4*, *7*, *9*
Bermuda fireworms, 38
Bermuda petrels (cahows, aka devil birds), 1, 6–10, 28, 241
blackcap warbler, 235–237, *236*
Blakemore, R., 113
bristle-thighed curlew, 186–187
butterflies, monarch, 19–20, 37, 40, 74, 105, 240

cahows (Bermuda petrels, aka devil birds), 1, 6–10, 28, 241
Canadian geese, 190
carbon dioxide, levels of, *229*
chemotaxis, 13
chimpanzees, 155–160
circadian rhythm, 40–45, *41*
Clark's nutcracker, 176–178
cliff swallows, 231–232
climate, global change in, 228–233, *229*, 237–241
clocks. *See* timers, period
clock-shift experiments: in ants, 73; in birds (other than pigeons), 87, 142, *143*; in homing pigeons, 134–137, *135*, *136*, 178, 197–198, *198*; in honey bees, 41
cognitive map, 32, 157–173
compass orientation and compasses: beacon based, 180–181; bearing based,

compass orientation and compasses (*cont.*)
169; definition of, 16; landmark based, 28–29, 96–100, *98*, *99*, 162–165, 166–169, 181, 190; lunar, 91; magnetic-field based, 100–115, 142–154; polarized-light based, 91–96, 138, *139*, 141, *141*; solar, 55, 65–67, *66*, 71–74, 82–91, *89*, *90*, 98–100, 134–137, 180; star (constellation) based, 138–141, *140*, 153
compound eyes, 74–77, *76*
conditioning, 157–158
conic map projection, 22–23, *23–24*
coordinates (radial, oblique, orthogonal), 194–195, *195*
copepods, 11–13, *12*
Coriolis force, 142–143
counting (as a way of measuring distance), 48–50
crabs, fiddler, 48–49, *49*
cranes, whooping, 231
cryptochrome, 110–111, *111*, 146–148
curlew, bristle-thighed, 186–187
cycles, annual (calendar), 57–68, 117; daily (circadian, clock), 40–45; lunar, 67–68, 70; tidal, 67–68, 69–70

dance language (of honey bees), 77–84, *79*, 88–90, *90*, 201–202
day length, 61–68, *63*, *66*
dead reckoning. *See* inertial navigation
deflector lofts, 202–203, *203*
desert ants, 50, 176
digger wasp, 173–175, *174*
displacement tests: of honey bees, 162–165, *164*, *165*; of humans, 223–225, *224*, *225*; of lobsters, 214, 216, *216*; of newts, 214–215, *215*; of reed warblers, 190, *191*; of sea turtles 216–222, *219*, *220–221*, *222*; of sparrows, 137, 192–193, *192*, 204. *See also* homing pigeons
distance, measurement of, 47–53, *49*, *53*
dogs, 155–156, *156*, 158
doves: mourning, 150; passenger pigeons, 150; rock, 120, 150. *See also* homing pigeon
drift, 5–6, 26–28, *27*, 30–31, 50–52, 129–130
Drosophila, 110, 147
dung beetles, 95–96

elasmobranchs, 147
Emlen, S., *119*, 140
entrainment (synchronization), 41–45, *41*, 57–68, 103, *104*
ethmoid sinus, 145–147
European robins, 117–118, 146–148, 212

fiddler crabs, 48–49, *49*
finches, 229–230; house, 234
fireworms, Bermuda, 38
flycatchers, pied, 146
flying squirrels, *41*
founder effect, 234
fruit fly, 110, 147

geese, Canadian, 190
geotaxis, 13
godwit, bar-tailed, 185–186, *186*
great circle, 23, *23–24*, 152, 194
great tits, 231
Griffin, D., 131
gyre, North Atlantic, 4–6, *4*, 217–222, *220–221*

habitat destruction, threat of, 240
hamsters, *163*
hippocampus, 172–173
homing pigeons: accuracy (precision) of, 123–127, *127*; baited-air tests with, 204, *205*; clock-shift tests of, 134–137, *135*, *136*, 178, 197–198, *198*; compass use by, 130, 134–137, 142–146, 149–150; deflector-loft tests of, 202–203, *203*;

evolution of, 120–121; first-flights vs. first-tests of, 126–127, 199; homing behavior of (generic), 122–127, *122*, *123*, *124*, *125*, *127*; inertial navigation by, 130; landmark use by, 125–126, 178; magnetic compass of, 143–145, *144*, *145*, 150; map sense of, 32, 189, 194–214; natural history of, 120–122, 126, 150; ontogeny of orientation in, 130, 136–137, 149–150, 189; release-site biases of, 206–208, *208*, *209*, 210, 212; strip-map use by, 178; sun compass use by, 134–137; tracks of, 125–126, *125*; use of palisade lofts in rearing of, 199–201, *200*

honey badgers (ratels), 170–171, *171*

honey bees: circadian rhythm of, 41–45, *45*; dance-language of, 77–84, *79*, 88–90, *90*, 201–202; detour experiments with, 167–168, *168*; displacement experiments with, 162–165, *164*, *165*; distance measurement by, 47–48, 50–52, *53*; evolution of, 35; eyes of, 74–77, *76*; foraging cycle of, 35–36, 44; interval timers of, 46–47; lake experiments with, 171–172, *172*; landmark learning by, 175–177, *175*, *177*; landmark use in sun compensation by, 97–100, *98*, *99*; local-area maps of, 162–165, *164*, *165*, 167–169, *168*; magnetic-field detection and use in, 100–105, *102*, *104*, 111–112, 114–115, 147; natural history of, 35–36, 84–85, 103–104; orientation under overcast/clouds by, 96–100, *98*, *99*; polarization compass of, 92, 95; speed measurement by, 50–53, *53*; sun compass of, 73–75, 80–86, 97–100, *98*, *99*, 180

honeyguides, 170–171, *171*

hormones, 117–118

house finches, 234

house wrens, 239

humans: distance measurement in, 48; local navigation in, 178–183, *179*, *180*, *182*; putative map sense of, 223–225, *224*, *225*

imprinting: definition of, 85–86; in navigation, 85–86, 126, 189, 218, 241–242

induction, 147

inertial navigation (aka dead reckoning), 5–6, 29–31, *30*, 129–131, 155–184; definition of, 16; in humans, 181–183; vs. piloting, 166–167

inner ear, 48, 130, 181, 213

intervals. *See* timers, interval

Kavanau, L., 160

Keeton, W. T., 143, 208

Kirschvink, J., 111

knot and nautical mile, definition of, 31

Köhler, W., 155–160

Kramer, G., 199–200; cage pioneered by, 118–119, *119*

lagena, 213

landmarks: learning of, 173–177, *174*, *175*, *177*; sun-compensation use of, 97–100, *98*. *See also* honey bees; map, ontogeny of formation (local-area); piloting

latent learning, 156–163, 160

latitude, 196, *196*

leeway, 26–28, 27, 31

Lindauer, M., 85

lobsters, spiny, 106, 214, 216, *216*

lofts: deflector, 202–203, *203*; palisade, 199–201, *200*

loggerhead sea turtles, 217–223, 219, 220–221, 222

Lohmann, K., 214, 216–223

longitude, 196–198, 196

loxodrome (aka rhumb line), 22–23, 23–24, 193

magnetic field: anomalies in, 208–211, *209*, *211*; calibration to, 149–153, 210–212; components of (total intensity, inclination, dip angle, vertical intensity, intensity-slope direction), *101*, 210; cryptochrome-based detection of, 110–111, *111*, 146–148; daily variation of, 103, *104*, 206–207, *207*; declination of, 105–106, *106*; detection of, 100–115, 107–115, 212–214 (*see also* magnetite; cryptochrome; induction); gradients in, 208–211, *209*, *211*; *212*; induction of, 102–103; paramagnetism, 107–110, *108*, *111*, 146–148; permanent magnetism (ferromagnetism), 107–109, *108*, 113–115; "storms" in, 206–207, *207*; superparamagnetism, 108–109, *108*, 111–112, *113*, 115; topography of, 208–211, *209*, *211*; *212*

magnetite, 13, 107, 112–114, 145–148, 212–213

map: bearing based, 169; cognitive (aka mental), 32, 157–173 (*see also* piloting); local-area (aka home-area, home-range), 162–173; ontogeny of formation (local-area), 167–169, 181; projections of, compared, 22–23, *23–24*; strip (aka sequential-landmark), 28–29, 168–169, 181, 190; true, 194–225

map sense, 185–225; vs. cognitive map, 32; coordinates of, 188–189, 194–195, *195*; definition of, 31–32; of humans, 223–225, *224*, *225*; logic of, 188; magnetic hypothesis of, 206–225; olfactory hypothesis of, 198–205, *200*, *203*, *205*

mazes and maze learning, 158–163, *160*, *161*, *163*

mean bearing, definition of, 122–123, *123*

Mercator map projection, 22–23, *23–24*, 152

mice, white-footed, 42, 160

migration: calibration in, 149–153; evolution of, 228–237; genetic basis of, 230–237; logic of, 59; long-distance, 230–231, 239; ontogeny in, 149–153; short-distance, 230–231, 233; strategies of, 131–132, 137, 229–242; in warblers, 235–237, *236*

migratory restlessness, 117–119

mole rats, 148, 212

monarch butterflies, 19–20, 37, 40, 74, 105, 240

Moore, B., 206

Morgan, C. L., and Morgan's Canon, 10–11

moths, 13–14, *15*, 106, 131

mourning doves, 150

nautical mile and knot, definition of, 31

navigation. *See* beacons; compass orientation; inertial navigation; piloting; taxis; true navigation; vector navigation

newts, 147, 214, *215*

nonsense orientation, 102–103, *102*

North Atlantic gyre, 4–6, *4*, 217–222, *220–221*

nutcracker, Clark's 176–178

Olton, D., 161–162

orientation. *See* compass orientation; inertial navigation; piloting; taxis; true navigation; vector navigation

palisade lofts, 199–201, *200*

Papi, F., 201–202

passenger pigeons, 150

periods. *See* timers, period

petrels, Bermuda (cahows, aka devil birds), 1, 6–10, 28, 241

phenotyope, plasticity of, 228–236, 239

Phillips, J., 202, 214

phonotaxis, 14–15

phototaxis, 12

phytoplankton, 11–13
pied flycatchers, 146
pigeons, homing. *See* homing pigeons
piloting (aka coastal navigation), 5, 24–28, *27*, 155–184; beacon-based, 180–181; bearing-based, 169; defined, 16; vs. inertial navigation, 166–167
plankton, 11–13
planning. *See* cognitive map
Polaris (North Star), 57–59, 138–139, *140*
polarization, 91–96, *92*, *93*, *94*, *95*
pole point, 138–141, *140*, *141*, 150–153, *151*

Randi, J., 223
ratels (honey badgers), 170–171, *171*
rats: lab, 158–162, *160*, *161*, 176; mole, 148, 212
rectangular map projection, 22–23, *23–24*
red-eyed vireos, 20–21, *21*
red-spotted newts, 214, *215*
reed warblers, 190–191, *191*, 233, *233*
release-site bias, 206–208, *208*, *209*, 210, 212
rhumb line (aka loxodrome), 22–23, *23–24*, 152, 193
right whales, 227
robins, European, 117–118, 146–148, 212
rock doves, 120, 150

salmon, 128, 147, 240–241, 242
San Antonio, 1, 3–7
sandhoppers, 69–70, 91
Sargasso Sea, 4–6, 217–222, *220–221*
Schmidt-Koenig, K., 197, 124
sea level, changes in, *229*, 237–238, *238*
seasons, source of, 59–62, *60*
sea turtles, 128, 147, 217–223, *219*, *220–221*, *222*, 241–242
seed caching, 176–178
selection, directional and normalizing, 231–235, *232*, *235*
sharks, 147
shorebirds, 185–187, 239, 240
sidereal day and year, 57–58, *58*
sounds, subsonic (infrasonic), 132–133
sparrows, *143*, 192–195, *192*, *193*, 204, 230, 239
speed, measurement of, 30–31, 50–53, *53*
spiders, salticid (aka jumping), 165–166
spiny lobsters, 106, 214, 216, *216*
squirrels, flying, *41*
strip map (aka sequential-landmark), 28–29, 168–169, 178, 181, 190
sun: arc of, 55, 65–67, *66*, 72–74, *72*; as a compass (other than compensation), 71–73, 134–137; compensation for movement of, 19–20, *20*, 72–74, *72*; 86–91, *89*, *90*, 98–100, 180; direction of movement of, 73–75, 82–86; identification of in bees, 80–82, *82*. *See also* clock shift experiments; polarization
swallows, bank, 198; cliff, 231–232
synchronization (entrainment), 41–45, *41*, 57–68

taxis (pl. taxes), 11–16, *14*, *15*, 19–20; definition of, 16
temperature, changes in, *229*, 233–234, *233*, 237–241, *238*
time: astronomical bases of, 54–68; measurement of, 35–68
timers: accuracy of interval timers, 46–47; accuracy of period timers, 43–45; distinguished, 37–39; entrainment of, 40–45, *41*, 57–68, 103, *104*; free-running period of, 40–42, *41*; interval (elapsed time, stopwatch), 46–47; period (clock, calendar), 40–45 (*see also* cycles)
Tinbergen, N., 173–175
tits, great, 231
Tolman, E., 158–162
tropocyte, 112

tuna, 147
true navigation, 31–32; definition of, 16. *See also* map sense
turbulence, 131
turtles, sea, 217–223, *219*, *220–221*, *222*, 241–242

vanishing bearing, definition of, 122–123, *123*
vector navigation, 3–5, 16, 20–22, *21*, 23, 193
vestibular system, 48, 130
vireos, red-eyed, 20–21, *21*
virtual-displacement tests: of lobsters, 214, 216, *216*; of newts, 214–215, *215*; of reed warblers, 190, *191*; of sea turtles, 216–222, *219*, *220–221*, *222*; of sparrows, 192–193, *192*, 204
visual flow, 47–53, *53*
von Frisch, K., 77–79, 86, 92, 97, 167–168
waggle dance (of honey bees), 77–84, *79*, 88–90, *90*, 201–202
Walcott, C., 143–145, 206, 208–210
Wallraff, H., 200
warblers: blackcap, 235–237, *236*; reed, 190–191, *191*, 233, *233*
warming, global, 228–233, *229*, 237–241
wasp, digger, 173–175, *174*
waterbirds, 190
Waterman, T., 47
whales, right, 227
white-crowned sparrows, 192–195, *192*, *193*, 204
white-footed mice, 42, 160
whooping cranes, 231
Wilcove, David, 241
Wiltschko, W. & R., 146–147
wrens, house, 239

zooplankton, 11–13, *12*
zugunruhe, 117–119